THE FOREST RANGER

A STUDY IN ADMINISTRATIVE BEHAVIOR

HERBERT KAUFMAN

SPECIAL REPRINT EDITION

RESOURCES FOR THE FUTURE / New York, NY, USA

An RFF Press book
Published by Resources for the Future

711 Third Avenue,
New York, NY
10017

2 Park Square, Milton Park,
Abingdon, Oxfordshire
OX14 4RN

Library of Congress Cataloging-in-Publication data

Kaufman, Herbert, 1922–
The forest ranger : a study in administrative behavior / Herbert Kaufman. – Special reprint ed.
p. cm.
Includes bibliographical references and index.
ISBN 1-933115-26-2 (hardcover : alk paper)
ISBN 1-933115-27-0 (pbk. : alk. paper)
1. United States. Forest Service. 2. Foresters–United States.
3. Industrial management–United States–Case studies. I. Title.
SD565.K3 2006
634.9′2092–dc22 2005029327

This book was typeset by TechBooks. Cover background, "Chalk Dust on Textured Paper," photographed by Siede Preis / Photodisc Green Collection / Getty Images.

The cover was designed by Maggie Powell.

ISBN 1-933115-26-2 (cloth) ISBN 1-933115-27-0 (paper)

About Resources for the Future *and* RFF Press

Resources for the Future (RFF) improves environmental and natural resource policymaking worldwide through independent social science research of the highest caliber. Founded in 1952, RFF pioneered the application of economics as a tool for developing more effective policy about the use and conservation of natural resources. Its scholars continue to employ social science methods to analyze critical issues concerning pollution control, energy policy, land and water use, hazardous waste, climate change, biodiversity, and the environmental challenges of developing countries.

RFF Press supports the mission of RFF by publishing book-length works that present a broad range of approaches to the study of natural resources and the environment. Its authors and editors include RFF staff, researchers from the larger academic and policy communities, and journalists. Audiences for publications by RFF Press include all of the participants in the policymaking process—scholars, the media, advocacy groups, NGOs, professionals in business and government, and the public.

About Resources for the Future and RFF Press

[illegible]

CONTENTS

Foreword

THE VAST FUNNEL

I chanced upon Herbert Kaufman's *The Forest Ranger* one day in 1961 in the halls at the University of Washington. I was on leave from the Forest Service to add a master's degree to my bachelor's in forestry, and one of my favorite professors, as we passed, handed me this new book about the agency.

I opened it, and in just the first few pages Kaufman showed me what I had not understood before about my employer: "As the lowest-ranking line executives and operators in national forest administration, [rangers] translate the words of policy statements—of federal statutes, departmental regulations, and Forest Service directives—into action" (p. 4). It wasn't the chief at the top, nor the 10 regional foresters, nor the 144 forest supervisors who ensured that the work on ground was the right job, done the right way. It was the district rangers, all 792 of them.

Kaufman's question was how an agency that was scattered among 792 administrative parcels—mostly in the West but also east of the Mississippi and in Alaska and Puerto Rico—could achieve what Congress and the executive branch asked of it. After the chief and his advisers interpreted departmental and congressional instructions, just how did word get passed down to the ground so that the myriad directives would be followed uniformly, and uniformly well? That is the nub of this fine book. Significantly, the view is from the bottom up, in line with the ranger's perspective: in the Forest Service, the administrative triangle is upside down, with the bureaucracy spreading out above and the point at the ranger's head.

Kaufman was a keen observer and a good listener as he gathered insights during week-long visits with 5 rangers. His subjects were all males and all forestry school graduates, as probably were the other 787 rangers. Back then, Forest Service administrative culture was uniform, from top to bottom. Ninety percent of professional employees were foresters, and the other 10 percent—engineers and social scientists—served as expert advisers.

There is good history in *The Forest Ranger*. In 1905, Chief Gifford Pinchot began establishing a system of decentralization: local issues would be dealt with locally. That quickly became the agency's mantra, and it remained virtually unchallenged for generations of rangers. Kaufman recognized that decentralization could add to "the tendencies toward fragmentation," but he saw that in the Forest Service, such tendencies were "not unchecked. They are overcome, or at least neutralized" (pp. 86–87). Most policies fit comfortably within the rangers' uniform culture, and it was the ranger who carried out the agency's mission on the ground.

Kaufman captured well the agency's effort to obtain statutory authority over logging on private lands—an effort that began in the 1920s and officially ended on January 20, 1953, with the inauguration of President Eisenhower. The chiefs had favored regulation, and carrying out contentious national policies at the local level was where the Forest Service excelled. But the Forest Service failed to gain regulatory authority. Instead, in later years, it would be the Environmental Protection Agency, the Fish and Wildlife Service, and the Corps of Engineers who would do the regulating, and the Forest Service became the regulated—subject to requirements for environmental impact statements, interdisciplinary planning teams, public participation in forest management planning, and compliance with legal mandates for clean water, clean air, and endangered species protection.

Much has changed since 1960, starting with the numbers. Budget cuts and consolidation made possible by improved communication and transportation have reduced the ranks of district rangers to (at last count) 522. American culture and values have changed dramatically over the past 45 years, and those of the Forest Service along with them. The agency no longer has an all-male cast of foresters; women and nonforesters now fill a large percentage of positions at all levels throughout the agency. In fact, one commonly hears that in the Forest Service today, forestry is no longer the lead profession, having been supplanted by biology and social science. The specialists who earlier would only advise the foresters are now in decision-making positions, in large numbers—causing a corresponding shift in the agency's approach to land management. And the agency's masters include not just the executive and legislative branches but also the judiciary, as today the Forest Service is besieged by litigation.

Yet Kaufman's work has remained highly relevant—to my own professional life as well as to students of organizations and administrative culture. In 1988, I began a long-term program to interview Forest Service chiefs after they retired, beginning with John R. McGuire. In preparation, I cribbed questions based on Kaufman's 1981 book, *The Administrative Behavior of Federal Bureau Chiefs*. Unlike the bottom-up look provided by *The Forest Ranger*, the later work looked at the qualities that made bureau chiefs successful. I was impressed that Chief McGuire had detailed knowledge of both books, which helped greatly as he described what it was like to be chief.

A decade later I interviewed former Chief Dale Robertson, the first chief to have been a forest ranger at the beginning of his long Forest Service tenure. Midway in his career, he took educational leave to earn a master's of public administration at American University. A professor assigned *The Forest Ranger*.

It was the late 1960s, and most of Robertson's young classmates were eager to challenge convention; products of their time, they thought the agency militaristic. Robertson realized that the Forest Service would soon be hiring from this new generation, and that program management would need to accommodate employees with different values.

Kaufman's administrative pyramid—"a vast funnel with the Ranger at the throat of it; all of the varied elements and specialties above him pour out materials which, mixed and blended by the Ranger, emerge in a stream of action in the field" (p. 68)—is still inverted. On occasion, in more recent times, the mix and blend of increasingly numerous and complex directives into a stream of actions could overwhelm a ranger.

Elapsed time allows us to see just how very accurate Kaufman's interpretations were, and today we find the Forest Service struggling to adapt to its very different operating environment. In his portrait of the proud, effective, and disciplined agency at midcentury, Kaufman provided a solid benchmark that enables us to take measure of the changes since the original publication of *The Forest Ranger*—the increased complexity of the mission, the litigation over forest plans, the splintering of a once-uniform culture and esprit de corps, the politicization of the agency and its often-contentious relations with Congress. Will the people and the forests be better off if the Forest Service continues to change, or is the agency losing important elements of its culture that ought to be preserved?

Harold K. Steen
Former President, Forest History Society
New Mexico State University
August 2005

FOREWORD

A GROUNDBREAKING, GROUND-LEVEL STUDY

It is rare that a book on American public administration approaches its 50th anniversary yet remains well known and widely appreciated for its lasting influence on the field. This is the case for *The Forest Ranger: A Study in Administrative Behavior*, by Herbert Kaufman. The work has had a profound impact; its previous birthdays have been recognized with reexaminations of its value. For this golden moment, it is being republished by Resources of the Future. The publisher reports that this is "by far" its best-selling title, and to this I say, deservedly so. Scholars of public administration have learned a great deal from Herbert Kaufman. A true understanding of administrative behavior in the execution (or lack of execution) of public policies requires just what Kaufman did 50 years ago in *The Forest Ranger*—diligent, sophisticated, detailed, even-handed, in-depth analysis moving up the governmental chain.

Few people can distinguish the forest from the trees better than Herbert Kaufman. Doing so wisely and in a way that is well grounded in operational realities requires taking into account all three values in Kaufman's framework of public administration: representativeness, neutral competence, and executive leadership. How was the professional competence of the rangers who worked in remote locations brought into play in the 1950s and harnessed to the agency's mission? How has the Forest Service been affected by representative societal values and how has this changed over time? How did its leaders perform 50 years ago, and how do they perform today in reconciling the swirling forces and factions of our Madisonian political system?

Kaufman's study of the Forest Service teaches us about the need to dig deeply to answer such questions. His book is a valuable resource and lively account for scholars and students interested in understanding the history and complexities (always complexities!) of American public administration. Taken together, the introductory material and Kaufman's Afterword for this edition provide the contemporary reader with a basis for understanding the Forest Service in mid-20th century and the organizational changes that have occurred since 1960 and their effects.

From the Afterword, we learn that the Forest Service became Kaufman's subject "serendipitously." His Columbia University dissertation focused on one forest district. Impressed by his work, Resources for the Future supported him to extend his study to five districts, focusing as he did in his dissertation on a bottom-up view of the degree to which, and the way in which, workers in the field responded to central directives. Kaufman's work was inspired and guided by Herbert Simon, whom he credits for "always focusing on how decisions are made."[1] While serving for a year in Washington as a management intern at the U.S. Bureau of the Budget, Kaufman was influenced by Simon, whose social science framework "provided a systematic way of thinking about my subject and opened new conceptual doors for me."[2]

Kaufman's later work looking at other public agencies asked similar questions about mechanisms of central direction and policy implementation.[3] As a lifelong student of this subject, he cares

[1] Phone interview with Herbert Kaufman, July 12, 2005.

[2] Herbert Kaufman, "Music of the Squares," *Public Administration Review* 56 (1996).

[3] See Herbert Kaufman, *The Administrative Behavior of Federal Bureau Chiefs* (Washington, DC: Brookings Institution, 1981); *Administrative Feedback* (Wash-

about the reasons why American political parties, for example, are so loosely controlled and other organizations (the military, the Forest Service) are so much more centrally directed and how this changes over time.[4]

Like any agency subject Kaufman could have happened upon, the Forest Service is distinctive. For one thing, it is a federal agency. Rangers are federal employees; their chief is appointed by the secretary of Agriculture. The fact that the Forest Service is a federal agency differentiates it from most agencies of domestic government, which are state or local. Even though the Forest Service stands out for this structure, this is not to shortchange Kaufman's role as our teacher. He deserves credit for his insistence that we dig down operationally to the front lines if we are to understand and assess the administrative behavior and effects of government policies that are supposed to influence society and the economy. *Implementation matters.*

Four years before *The Forest Ranger* was issued, Kaufman published a seminal article in *The American Political Science Review,* "Emerging Conflicts in the Doctrines of Public Administration," which summarized the three values of American public administration mentioned earlier. Kaufman wrote, "The central thesis of this paper is that an examination of the administrative institutions of this country suggests that they have been organized and operated in pursuit successively of three values, here

ington, DC: Brookings Institution, 1973); *Are Government Organizations Immortal?* (Washington, DC: Brookings Institution, 1976); *The Limits of Organizational Change* (Tuscaloosa: University of Alabama Press, 1971); *Red Tape* (Washington, DC: Brookings Institution, 1977); and *Time, Chance, and Organizations* (Chatham, MA: Chatham House Publishers, 1991). See also Eugene Bardach, *The Implementation Game* (Cambridge, MA: MIT Press, 1977); Richard F. Elmore, "Organizational Models of Social Program Implementation," *Public Policy* 2 (Spring 1979), pp. 186–228; Irene Lurie, "Field Network Studies" in *Policy in Action: Implementation Research and Welfare Reform,* Mary Clare Lennon and Thomas Corbett, eds. (Washington, DC: Urban Institute Press, 2003); and Richard P. Nathan, *Social Science in Government* (Albany: Rockefeller Institute Press, 2000, 2nd edition).

[4] Phone interview.

designated representativeness, neutral competence, and executive leadership."[5]

Applying this framework, Kaufman says in his Afterword to this edition of *The Forest Ranger* that new forces impinging on the agency have made its job more challenging and the task of resisting centrifugal forces more difficult. He cites environmental, recreational, and civil rights policies that have diluted the solidarity of mission of the Forest Service. "Significant and rapid changes in the Forest Service's environment did occur after World War II, and by the time the agency completed the first half-century of its existence in 1955, its preeminence in the conservation movement had begun to erode."[6] Furthermore,

> Meanwhile, the civil rights movement was making headway. Laws protecting the rights of women and minorities were added to the statute books, new administrative agencies were created to enforce these provisions, and new court decisions backed them up. Longstanding staffing practices would no longer pass muster.[7]

Go back now to a point made earlier that the Forest Service is not your typical government agency. (Indeed, "typical" is a very hard term to apply when studying American public administration.) Many national domestic policies operate by *indirection* in American federalism. The roles and structures of state and local public agencies are highly varied. The federal government seeks to influence state and local agencies in several ways—by preemption, by mandates, and more subtly by giving them money (grants-in-aid) and setting conditions for the use of these funds. State and local public agencies must contend not only with Washington but also with their own governments and with myriad other public agencies and private and nonprofit organizations and interest groups that care about and have power in carrying out (or

[5] Herbert Kaufman, "Emerging Conflicts in the Doctrines of Public Administration," *The American Political Science Review* 50(4) (December 1956), pp. 1057–73.

[6] Herbert Kaufman, Afterword, p. 264.

[7] Ibid., p. 264.

impeding) the purposes that they are supposed to achieve. This is not meant as a complaint. We are blessed to have multiple access points and a yeasty democratic form.

Other ground-level studies of the administrative behavior of government agencies have been published, but they tend to be linked to different functional areas of government. In 1980, 20 years after the publication of the *Forest Ranger*, Michael Lipsky published *Street Level Bureaucracy: The Dilemmas of Individual Public Services.* Lipsky observed local bureaucrats in schools, police and welfare departments, lower courts, and legal services to learn about "how the rules are experienced by workers in the organization and to what other pressures they are subject."[8] Another important book in this literature was written by Kaufman's student, Aaron Wildavsky, who along with Jeffrey Pressman and Angela Brown published several editions of the well-known book *Implementation,* a study of the U.S. Economic Development Administration's employment efforts in Oakland, California, in the 1960s.[9]

James Q. Wilson's book *Bureaucracy*, based on his lectures at Harvard, is another classic in the field. Wilson specifically comments on the influence of *The Forest Ranger*, calling it "Perhaps the best-known study of a government agency trying to manage the behavior of operatives working in remote places." He draws special attention to Kaufman's description of the reporting requirements for forest rangers and the arrangement whereby rangers were moved regularly from place to place. Let's let Kaufman's tone and style in *The Forest Ranger* speak to these two points. On reporting requirements, he had this to say:

> In the Forest Service, requests for reports invariably specify in detail the exact information desired, the format in which it is to be presented, the period it is to cover, the dates in which it is to be filed, and the place to which it must be sent. As a rule, reports are submitted

[8] Michael Lipsky, *Street-Level Bureaucracy: Dilemmas of the Individual in Public Services* (New York: Russell Sage Foundation, 1980), p. xi.

[9] Jeffrey L. Pressman and Aaron Wildavsky, *Implementation* (Berkeley: University of California Press, 1984).

> on prepared forms, which thus prescribe precisely what information must be furnished and how it shall be organized.[10]

In a similar vein, Kaufman saw the 1950s Forest Service policy of frequently transferring rangers "as a means of inducing men to conform and of exposing nonconformance," which, he said, "exerts constant integrative pressure."[11]

> The Rangers studied here have had differing experiences, but all have employment records that reflect the general statements of transfer policy. Of three with more than 20 years of the Forest Service, one was in five locations, one in four, and the third in three within a dozen years; and each moved again at least once later on. As for the younger men, one has served in four places in five years, the other in two places in seven years. Three served in one capacity or another on the staffs of forest supervisors as well as at the district level.[12]

Commenting on the Forest Service's "high degree of unity," Kaufman asked what would happen in the future. Victory is "never won once and for all."[13] Mulling the "hazards of success," Kaufman saw then, and still sees, a need to study administrative behavior over time in order to probe the ways in which organizations can overdo central direction and what the consequences are when they do so. He was, and still is, fascinated by the evolution of organizations. He wanted others to follow up on his question about the hazards of administrative success—the possibility that a tightly run organization like the Forest Service that is well adjusted to its environment could implode if the environment changed.

As it has turned out, the character and role of the Forest Service have changed, but not fundamentally. Kaufman hoped other experts would revisit this territory, digging as deeply as he did in conversations with the rangers and their staff. Although there has not been as thorough a followup as Kaufman had wished for,

[10] Herbert Kaufman, *The Forest Ranger: A Study in Administrative Behavior* (Baltimore: Johns Hopkins Press, 1960), pp. 126–27.
[11] Ibid., p. 156.
[12] Ibid., p. 177.
[13] Ibid., p. 207.

Harold K. Steen's histories and Kaufman's own Afterword to this edition provide a basis for considering the kinds of forces and factors that can cause institutional changes in the way front-line officials receive and respond to policy signals. According to Kaufman, the agency's response to these challenges, "albeit gradual and unenthusiastic," confirm his expectation that adjustments would come, but not easily.

Kaufman is impressed that of all his many works, *The Forest Ranger* is still remembered particularly well. He says his mentor at the Institute of Public Administration, Luther Gulick, told him that if he clearly and patiently described the ground-level behavior of a public agency, he would be cited everywhere—in modern parlance, that his book would have legs. Gulick was right. There is still a need for this kind of deep digging about organizational behavior, particularly in ways that take advantage of data-handling technologies now available and blend effective narrative with statistical analysis about the crucial governmental challenges involved in turning promises into performance. This is the shadow land of public administration that Herbert Kaufman taught us how to illuminate. As T.S. Eliot said in "The Hollow Man,"

Between the idea
And the reality
Between the motion
And the act
Falls the Shadow

Richard P. Nathan
Co-director, Nelson A. Rockefeller Institute of Government
State University of New York, Albany
August 2005

THE FOREST RANGER:

A STUDY IN ADMINISTRATIVE BEHAVIOR

[Original 1960 Edition]

The charts and map in this book are reproduced through the courtesy of the Forest Service, United States Department of Agriculture.

FOREWORD

Resources for the Future, after having given financial support to the preparation of this book, is glad to add it to its series of studies on natural resource use and management. We believe that it is a real contribution to knowledge about public resource management.

The public programs in resource management are important because of their scope, and hence the ways in which they are carried out are of concern to all students of resource management. The federal government owns 24 per cent of the land area of the United States, if the Indian and military lands are included, and 19 per cent if they are excluded. More than half of the standing saw-timber of the nation is on these lands (including the Indian reservations). The demand for the various products of the federal land is increasing rapidly from year to year, and both the expenditures on such lands and the revenues from them are rising steadily. The federal lands of our nation are too extensive and too important for us to tolerate administration or management at less than the full economic potential. It was a consideration of these facts that led Burnell Held and me to write *The Federal Lands: Their Use and Management* a few years ago. In that book, we present a great deal of information about the federal land programs.

In addition, the federal government conducts many water development programs which present some of the same administrative problems. There are also state land management programs, in some states, with many comparable features. Even the larger private land owning and land managing companies encounter similar problems.

Most of the literature on public and private resource management is concerned with economic problems and public policy—that is, with analysis of the relative profitability of different programs, or with their relevance to various social goals. It is often assumed, explicitly or subconsciously, that the goals of the top echelons of the organization will be translated more or less automatically into specific actions at the local level. But this may not be true; certainly it should not be assumed to be true, without the most careful examination of the facts. Neither the researcher, intent on understanding how a program actually works, nor an administrator determined to achieve certain goals, should assume that what is decided at the top is the same as what is done at the bottom of the organization.

In the management of natural resources, it is the man on the ground who actually carries out the program. This is equally true for private and public organizations. It is what he *does*, not what the department secretary, bureau chief, or company president *says*, that actually makes the program. If a resource management program is well conceived, its various specialized parts must come together into a rational whole at the very top of the organization, and also at the lowest operating level. Specialists at various intermediary levels may look at only those aspects of the total program in which they are interested. But the forest ranger, or the comparable local official in other agencies, must somehow weave the various specialized programs into a whole, that makes sense to him, his superiors, and his public.

In public administration, it is all too common to look at agency organization from the top down. Organization charts and description start with the head, then go to the branches, and finally to the lower levels. There is discussion of such matters as delegation of authority and of control by the top. This is all important; but perhaps more significant is to look at the organization upwards from the lowest general purpose unit—in this case, from the ranger district upward. This is what the present study has done.

Federal resource management programs are inevitably spread over wide areas of the nation and its possessions. They must operate under a wide variety of conditions, natural, economic, and cultural. If the local offices are to be close to the resources and to the users, there must inevitably be a considerable number of such offices, often with only a few men in each. The men in charge of local areas and of these offices must be empowered to act, at least within defined limits. When there is a forest fire, the man on the ground must take immediate charge; and when there is an application for a timber sale, he must know how to proceed expeditiously. The central office may guide him in many various ways, but it must give him authority to act if he is to do a good job. It cannot supervise his day-to-day actions.

Out of this physical situation comes the major dilemma of the central office of a federal resource managing agency: how to devise and operate an agency which will operate consistently, in the sense of reducing to the minimum variations from established, organization-wide norms while at the same time preserving individuality and stimulating creative thinking and action on the part of its men. As far as reasonably practical, an agency will seek to have the actions taken by one man the same as those which would have been taken by any other man in the organization in the same circumstances. This is the purpose of regulations, procedure manuals, and the like. At the same time, the many and difficult problems of resource management cannot be reduced to rote; there is always the need for judgment and skill on the part of the individual. This is a complex problem, with many facets.

The United States Forest Service has had long experience in dealing with precisely this problem. Its present methods of administration are the outgrowth of trial and error over several decades. Most students of resource management and of public administration will agree that, on the whole, it has done an outstanding job. This book presents a detailed analysis of the operations of this agency, as viewed from the standpoint of the district ranger.

The ideas advanced in this book are Dr. Kaufman's. He neither sought our approval of his findings, nor did we require it. We do endorse the book as a thoughtful and imaginative study, worthy of attention.

Marion Clawson, *Director*
Land Use and Management Program
Resources for the Future, Inc.

September 1, 1959

PREFACE

This is not, in a strict sense, a book about federal forest policy. Except incidentally, it does not deal with the processes in which such policy is formed, or with the desirability or defects of prevailing policy, or with measures that might improve policy or the methods of policy formation. In a general way, current policy is taken as a "given."

Nor is this, in a strict sense, a book about the political life of the United States Forest Service—about its dealings with Congress, its position as a component of the Department of Agriculture, its rivalries and accommodations with other departments and bureaus, its negotiations and battles and adjustments and alliances with interest groups, its relations with political parties, its co-operation with state land management agencies and other state and local governmental organizations, its role as a Presidential instrument, its protection (and expansion) of its jurisdiction and power, or its strategic compromises with the forces in its environment.

This is not even a book about "bureaucratic politics"—that is, about the competition and maneuvers and intrigues of individuals jockeying to insinuate themselves into leadership positions.

These aspects of organizational life have been excluded

from the scope of this volume not because they are less important than the problem to which the volume is addressed, which is certainly not true, but because they have been more thoroughly explored. Intrinsically dramatic and obviously significant, they are the attributes of organization with which much of the writing on organization has been concerned. Generally, however, such analyses tend to take it for granted that a policy decision reached at a high level, an accommodation arrived at by agency leaders, or the rise to power of a new leader or group of leaders will somehow result in a modification of the behavior of the lower echelons—the people who do the physical work of the organization—so that the high-level understandings are converted into appropriate tangible accomplishments. A tacit assumption in much administrative literature is that what the top officers of an organization want, the organization does, and that this is too routine to warrant study.

Perhaps this is frequently the case. But often it is not, as any military commander whose troops have broken and run from fire, or any labor union leader who has been embarrassed by a wildcat strike, or any department head whose program has been sabotaged by a recalcitrant bureau chief, or any law enforcement officer whose subordinates have been found guilty of corruption, or any of dozens of other kinds of "leaders" chagrined to discover their "followers" are no longer following, can sadly testify. It does not "just happen" that the daily decisions and actions of the lower echelons make concrete realities of policy statements and declared objectives of the leadership; this takes planning and work.

The questions to which this book is addressed are: What kind of planning? What kind of work? The evidence indicates the Forest Service has enjoyed a substantial degree of success in producing field behavior consistent with headquarters directives and suggestions. How has this been accomplished? What are the practices and procedures by which such results are achieved?

For these purposes, it makes no difference whether what is done by the Forest Service is (as its supporters maintain) or is not (as its critics sometimes contend) what it *should* be doing. All that is relevant here is that the men in the field are apparently doing what the top officers want done in the field; the study aims at explaining how the wishes of the latter are transformed into the actions of the former.

Nor are the relations between Forest Service headquarters and the environment of the Service relevant here. Headquarters personnel may strike bargains, make concessions, press forward or retreat. What this study is concerned with is the way the field men are induced to carry out tangibly the terms of headquarters agreements.

Moreover, even the strategies pursued by eligibles in quest of the top positions in the Forest Service have only remote bearing on the central questions of this study. By and large, the men in the field seem to follow easily the men who attain the official positions of authority.

Of course, substantive policy problems, techniques of organizational survival, and bureaucratic politics are not completely separable from the problem of influencing field behavior. But they are peripheral to it, and the reader concerned with them alone will find little of interest in the pages that follow.

It is the reader curious about the often neglected relationship between the broad pronouncements at the top levels and the day-to-day activities of those who perform the physical tasks of an agency for whom this volume may hold some interest. Field compliance in the Forest Service is not total, naturally, but it is so high, despite powerful factors tending to reduce compliance, that it cries out for study. The effort to describe the way it is obtained is essentially an effort to capture that elusive phenomenon, "what actually happens in an organization."

In a manner of speaking, there is nothing more to this process than telling subordinates what to do and seeing that they do it. In a large-scale organization, however, this proves to be far from simple. And that is why it takes a book to try to explain it.

Herbert Kaufman
Department of Political Science
Yale University

June 1, 1959

ACKNOWLEDGMENTS

So many people have contributed to the preparation of this volume that it is almost inappropriate that the name of only one of us should appear as its author. Had the original pilot research not been supported by the Institute of Public Administration, the book probably would not have been written at all. Nor could it have been done had not the field research and the preparation of the manuscript been encouraged and sustained by Resources for the Future. Above all, it would have been impossible without the wholehearted co-operation of the United States Forest Service. Conceivably, some of these institutional co-operators may wish to dissociate themselves from much—or even all—of what I have written, but I am afraid they cannot escape the fact that the existence of this volume owes as much to them as it does to me.

I am particularly grateful to Luther Gulick, of the Institute of Public Administration, and Marion Clawson, of Resources for the Future. As for the Forest Service, I am in profound debt to so many people—for warm and gracious hospitality as well as for unstinting contributions of time and information and wisdom—that a full listing of these obligations would be excessively long and any selection might seem arbitrary and invidious. Yet I think all in the Service who helped me will understand if I express my appreciation to them in the form of deep thanks to the men in the field, with whom I spent most of my time—Clarence E. Anderson, Conrad W. Carlson, Richard E. Elliott, Rexford A. Resler, Chester A. Shields, and Fred J. Wirth—and to the members of the Washington office who made the field work possible—Lyle F. Watts (now retired), the late Earl W. Loveridge, Richard E. McArdle, Clare Hendee, and Gordon W. Fox.

The manuscript made possible by the generosity of those mentioned above would suffer from even greater and more numerous shortcomings than it now does if I had not had the benefit of advice and criticism from a great many people who were kind enough to take the time and trouble to read and comment upon the first draft or upon the original pilot study report. James W. Fesler, Charles McKinley, Charles H. Stoddard, and Robert Wolf took extraordinary pains in going over the manuscript in great detail. I also drew heavily on the ideas and suggestions of Herbert A. Simon, whose *Administrative Behavior* provided much of the initial stimulus for this volume. And I benefited greatly from the comments of Arthur W. Macmahon, Grant McConnell, Walter H. Meyer, John D. Millett, Wallace S. Sayre, Melvin Thorner, Norman I. Wengert, Albert C. Worrell, Robert J. M. Matteson, and Henry Jarrett. In addition, twenty-five copies of the manuscript were circulated among members of the Forest Service, who saved me from a great many technical errors.

I unquestionably would have bogged down in the burdens of correspondence, filing, transcription of field notes, and other administrative chores save for the talent of Florence Perry, who handled all these with efficiency and dispatch, and also typed most of the manuscript. Peter H. Seed gave invaluable help in getting the manuscript in final shape. Nora Avins typed parts of the manuscript, and Violet Stevens and Eleanor Wheeler typed the results of the pilot study.

This book is thus a joint endeavor. The principal justification for attributing it to me alone is that none of my collaborators sought to do more than aid and advise me; the last say about the organization, interpretation, and presentation of the material was left to me. I have followed the facts wherever they seemed to me to lead, even when I was thereby compelled to modify or reject suggestions. In this respect, this book is a personal, an individual, responsibility. This responsibility no independent researcher can escape.

New Haven, 1959 H. K.

[illegible] for [illegible] it [illegible] that [illegible] of my [illegible] might do [illegible] that and [illegible] the [illegible] about the organization, interpretation, and presenta- tion [illegible] I have followed the facts [illegible] me to lead, even when I was [illegible] In this respect, this book is a [illegible] individual [illegible] responsibility. This [illegible] can escape.

[illegible] H. S.

INTRODUCTION

CHAPTER 1

SUBJECT AND APPROACH

Policy is enunciated in rhetoric; it is realized in action. The creators and managers of organizations formulate and promulgate policy pronouncements, but, as Simon has pointed out, "The actual physical task of carrying out an organization's objectives falls to the persons at the lowest level of the administrative hierarchy."[1] Hence,

> In the study of organization, the operative employee must be at the focus of attention, for the success of the structure will be judged by his performance within it. Insight into the structure and function of an organization can best be gained by

[1] "The automobile, as a physical object, is built not by the engineer or the executive, but by the mechanic on the assembly line. The fire is extinguished, not by the fire chief or captain, but by the team of firemen who play a hose on the blaze. . . . It is the machine gunner and not the major who fights battles . . ." H. A. Simon, *Administrative Behavior* (New York: The Macmillan Company, 1947), pp. 2-3.

See also L. Gulick, "Politics, Administration, and the 'New Deal,'" *Annals of the American Academy of Political and Social Science*, Vol. 169 (September 1933), p. 62: "Much of the actual discretion used in administration is used at the very bottom of the hierarchy, where public servants touch the public." And C. I. Barnard, *The Functions of the Executive* (Cambridge: Harvard University Press, 1938), p. 139: "The individual is always the basic strategic factor in organization. Regardless of his history or his personal obligations he must be induced to cooperate, or there can be no cooperation."

> analyzing the manner in which the decisions and behavior of such employees are influenced within and by the organization.[2]

In the United States Forest Service (a bureau of the United States Department of Agriculture), these employees are the district Rangers. There are 792 Rangers in the Forest Service —not, as many people apparently suppose, lookouts sitting in lonely towers on isolated mountain peaks and watching for fires, or police officers like the members of the famous Texas constabulary bearing the same designation, but land managers in command of Ranger districts, the smallest territorial and administrative subdivisions of the national forests. They are charged with the protection, proper use, and development of all the resources in their districts and, while they have employees to help them in many of their tasks, they perform a great many themselves or personally oversee their subordinates; most of them spend more than half their time in the woods. Thus, while they are supervisors, they are also field workers. As the lowest-ranking line executives and operators in national forest administration, they translate the words of policy statements—of federal statutes, departmental regulations, and Forest Service directives—into action. The purpose of this study is to analyze the way their decisions and behavior are influenced within and by the Service.

There are powerful centrifugal forces that might be expected to fragment the Forest Service, to make each Ranger district a small, relatively autonomous Forest Service by itself, so that the appearance of 792 different policies in practice might seem no less likely than the emergence of a common policy for all. This fragmentation does not occur. The field men, it will be seen later, are guided extensively in their day-to-day actions by the leaders of the Forest Service. Rarely does one hear it said that Rangers behave in a fashion inconsistent with Service policy; on the contrary,

[2] Simon, *op. cit.*, p. 3.

they are sometimes described as too zealous in their conformance. Why this happens, why they do not succumb to the centrifugal forces inherent in the administrative situation, are the questions to which this volume is addressed.[3]

This focus of attention excludes two important considerations about the activities of public administrative agencies. One is an evaluation of the wisdom, the soundness, of the policies pursued by the Forest Service. The other is the "external" relations of the agency leaders with Congress, with the President, with other line and staff agencies, with interest groups (both friendly and hostile), with political parties and party functionaries, with state and local governments and agencies. Both issues are unquestionably extremely significant, but they are not relevant to the questions this study was designed to answer. For field forces may execute to the letter policies widely regarded as unsound, or, by the same token, may, by departing from their instructions, upset strategies deemed rational by many observers; whether or not their actions are congruent with policy does not necessarily depend on the goodness or badness, the rightness or wrongness, of the policy. And when adroit handling of their foreign relations by the leaders of the Forest Service ensure the agency's jurisdiction, position, funds, and status—when its survival is thus guaranteed for a time—the problem of unifying the activities of the men in the field remains; insofar as Service relations with the environment impinge on the internal functioning of the organization, they are examined here, but they are extraneous for the most part and therefore are not treated at length. The Rangers at the base of the administrative pyramid are subjected to influences pulling and pushing them in many different directions, yet they are held together; the restricted but difficult goal of this report

[3] This approach is also urged in E. O. Stene, "Administrative Integration," *Annals of the American Academy of Political and Social Science*, Vol. 292 (March 1954), pp. 111-19.

is to explain this integration by identifying and describing the factors at work and the interplay among these factors.

To understand the administrative behavior of the Rangers requires a thorough knowledge of the organization levels above them; to quote Simon again,

> Persons above this lowest or operative level in the administrative hierarchy are not mere surplus baggage . . . Even though, as far as physical cause and effect are concerned, it is the machine gunner and not the major who fights the battles, the major is likely to have a greater influence upon the outcome of a battle than any single machine gunner.[4]

Studying the Rangers therefore means studying the whole hierarchy.

To understand the administrative behavior of the Rangers also requires study of the history of the Forest Service. The members of all organizations are governed by values, beliefs, and customs that are almost indiscernible if research is confined to short periods in the evolution of the organizations. Time is a factor to be reckoned with.

Finally, understanding the Rangers necessitates a familiarity with the context of federal laws, regulations, and procedures in which they, as public servants, operate. The Forest Service is implanted in a larger bureaucratic setting from which it is inseparable and which accounts for many of its characteristics. The environment must be deliberately explored.

So the method of inquiry of this study involves the standard techniques of the political scientist—examining administrative structure, interviewing top-level officials, compiling and interpreting organization charts, scrutinizing statutes and government-wide practices, reading history and assembling chronologies, perusing manuals and hearings and reports, gathering evaluations and criticisms of the agency, and other-

[4] Simon, *op. cit.*, pp. 2-3.

wise directing attention to the context of the specific problem.

But a study concerned primarily with field officers presupposes much more than this. The traditional data are of interest only insofar as they are influences on the decisions of the field men. Intensive field investigation is needed to assess their impact and identify other influences that reinforce or counterbalance them. For this, the approach of the cultural anthropologist furnishes a most useful model:

> The field is the laboratory of the cultural anthropologist. To carry on his field-work, he goes to the people he has elected to study, listening to their conversation, visiting their homes, attending their rites, observing their customary behavior, questioning them about their traditions as he probes their way of life to attain a rounded view of their culture or to analyze some special aspect of it.[5]

This is the approach employed in this study of the Forest Service. That is not to say this is a work in cultural anthropology; it would be exceedingly presumptuous for anyone not rigorously trained in that discipline to claim to have mastered its methods, let alone to apply those methods to a type of social organization that only a few skilled anthropologists have ventured to investigate. Moreover, anthropologists in the field customarily allot time for far more extended sojourns with their subjects than was possible in this inquiry. Nevertheless, their practice of immersing themselves in the culture or subculture they propose to describe suggested a means of gathering the kind of information necessary to answer the questions about administrative behavior posed here. Within the limits of time and resources available, and recognizing the dangers of confusing superficially similar research activities with identity of method, the researcher adopted a pattern of data gathering that followed, albeit distantly, the anthropological pattern.

[5] M. Herskovits, *Man and His Works* (New York: Alfred A. Knopf, 1949), p. 79.

The Rangers and Their Districts

Five Ranger districts in different parts of the country were selected for intensive analysis. They were not chosen as being "typical," although districts with conspicuously unusual characteristics were avoided; rather, they were picked as samples because together they show almost the whole range of Forest Service activities and a wide variety of the conditions under which its work is carried on. Ideally, the number of "specimens" would have been larger—at least one district of a national forest in each of the ten regions of the Forest Service—but there was neither sufficient money nor manpower for more inclusive coverage. Conceivably, the available resources might profitably have been concentrated on two or three of the districts, permitting greater depth in the research, but this would have run the risk of getting distorted results because of conditions or procedures peculiar to individual areas, or because of the unique biases or traits of individual Rangers; with five districts, the likelihood that unique characteristics would cancel out appeared considerably greater. In short, this seemed to be the smallest number that could be studied without rendering all the findings seriously suspect, and the largest number available time and finances would support.

DISTRICT NUMBER ONE

One district is in Region Seven, the Eastern Region. Region Seven takes in fourteen states from Maine to Kentucky, but it is not large from an administrative point of view, for it contains fewer national forests (seven) than any other region in continental United States, the smallest acreage (a little over 4 million acres owned by the federal government), and has the smallest revenues, expenditures, and staff. The paucity of federal holdings in this area reflects

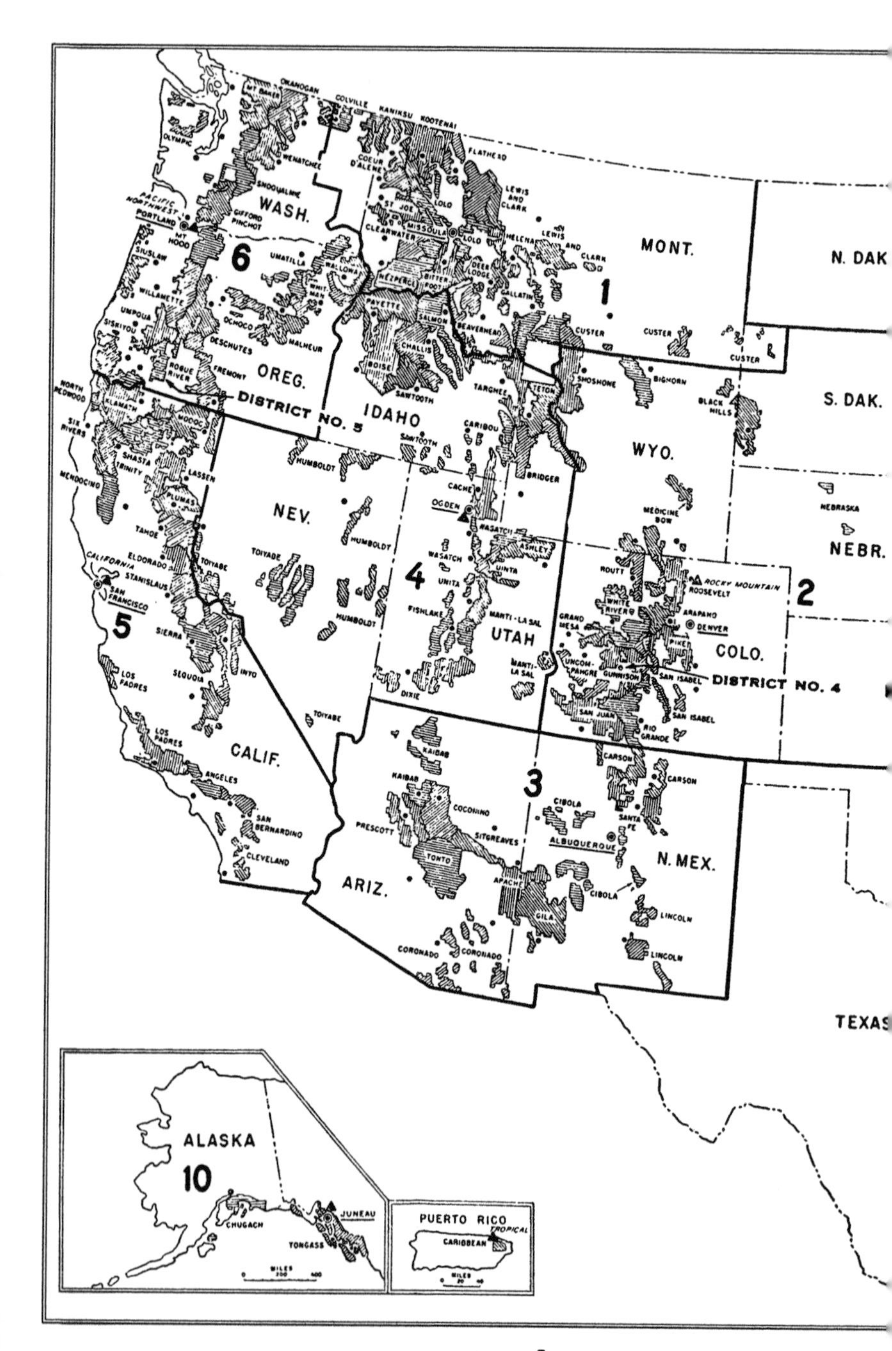
WASH.
OREG.
IDAHO
MONT.
N. DAK
S. DAK.
WYO.
NEBR.
NEV.
UTAH
COLO.
CALIF.
ARIZ.
N. MEX.
TEXAS
DISTRICT NO. 5
DISTRICT NO. 4
ALASKA
10
PUERTO RICO
CARIBBEAN
JUNEAU
CHUGACH
TONGASS
PORTLAND
MISSOULA
OGDEN
DENVER
ALBUQUERQUE
SAN FRANCISCO

FIGURE 1

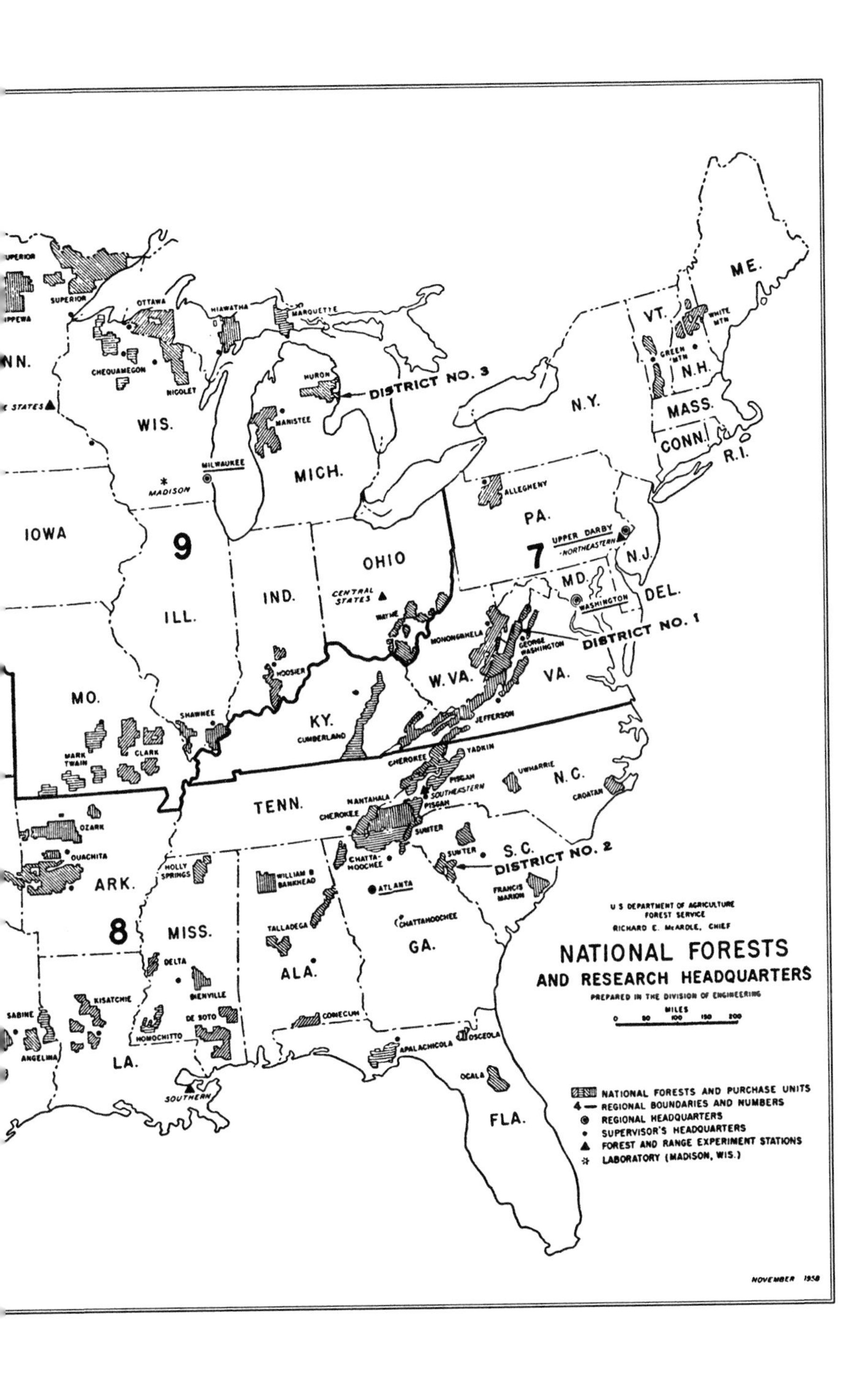

SUPERIOR
SUPERIOR
CHIPPEWA
MINN.
OTTAWA
CHEQUAMEGON
NICOLET
HIAWATHA
MARQUETTE
HURON
DISTRICT NO. 3
MANISTEE
WIS.
MICH.
MILWAUKEE
MADISON
IOWA
9
ILL.
IND.
OHIO
CENTRAL STATES
WAYNE
HOOSIER
MONONGAHELA
GEORGE WASHINGTON
DISTRICT NO. 1
W. VA.
VA.
JEFFERSON
KY.
CUMBERLAND
SHAWNEE
MO.
MARK TWAIN
CLARK
N.Y.
ALLEGHENY
PA.
7
UPPER DARBY
NORTHEASTERN
N.J.
MD.
WASHINGTON
DEL.
VT.
GREEN MTN
WHITE MTN
N.H.
ME.
MASS.
CONN.
R.I.
CHEROKEE
YADKIN
PISGAH
SOUTHEASTERN
UWHARRIE
N.C.
CROATAN
NANTAHALA
CHEROKEE
PISGAH
TENN.
SUMTER
SUMTER
S.C.
DISTRICT NO. 2
FRANCIS MARION
CHATTAHOOCHEE
ATLANTA
CHATTAHOOCHEE
GA.
OZARK
OUACHITA
ARK.
8
HOLLY SPRINGS
MISS.
WILLIAM B BANKHEAD
TALLADEGA
ALA.
DELTA
BIENVILLE
DE SOTO
HOMOCHITTO
KISATCHIE
SABINE
ANGELINA
LA.
SOUTHERN
CONECUH
APALACHICOLA
OSCEOLA
OCALA
FLA.
U S DEPARTMENT OF AGRICULTURE
FOREST SERVICE
RICHARD E. McARDLE, CHIEF
NATIONAL FORESTS
AND RESEARCH HEADQUARTERS
PREPARED IN THE DIVISION OF ENGINEERING
MILES
0 50 100 150 200
NATIONAL FORESTS AND PURCHASE UNITS
4 — REGIONAL BOUNDARIES AND NUMBERS
REGIONAL HEADQUARTERS
SUPERVISOR'S HEADQUARTERS
FOREST AND RANGE EXPERIMENT STATIONS
LABORATORY (MADISON, WIS.)
NOVEMBER 1958

the fact that this section of the country was the first to be settled, was completely in private hands by the time the Forest Service was established, and was not opened to Forest Service management until legislation permitting the purchase of lands for the national forests was enacted in 1911; the amount of land the government was able to buy was limited by the private demand for property in these heavily populated areas.

Originally, much of the land acquired by the government had been cut and burned over and farmed out until it was no longer capable of producing commercial crops, and it lay idle, unprotected, and in tax default; stripped of its cover, it deteriorated further because of erosion and uncontrolled fires. During the thirties, good, productive land also was placed on the market because of the depression, so government purchases added to the national forests of the region many fertile acres for silviculture. The entry of the Forest Service promised to return uneconomic land to eventual economic utility, so the purchases were welcomed by the states and communities affected.

For a long time, Region Seven was engaged primarily in a holding operation, making as much use of the resources as was feasible, but depending on natural processes and control of fire and disease and erosion to restore the productivity of the abused sections and on the recovery of the economy to restore the market for the productive sections. Since the end of the Second World War, there has been great progress in both respects, with a consequent shift from custodial administration to resource management. Management of timber and control of fire are the chief activities, but recreation and wildlife protection have arisen sharply in importance in recent years.

The Ranger district studied is the largest of six on the largest national forest in Region Seven. Comprising a net acreage (i.e., acreage actually owned by the federal government, and excluding enclaves of private property within the gross boundaries) of more than 237,000 acres, it lies mostly in Virginia, but some 10 per cent of it extends over the Shenandoah Mountains into West Virginia. With Washington, D. C., a scant 125 miles to the northeast, and a main highway running through the area from Washington to Roanoke and other southern points, the numbers of visitors and travelers continue to mount rapidly. Hunting and fishing are popular sports on the district, and there is a great deal of picnicking and some camping. The fire problem has increased in intensity as a result. Commercial harvest of timber is now substantial, with small transactions gradually giving way to large ones of millions of board feet. Water for two cities and many smaller towns and communities originates on the district, and some of it is stored on the district. The district is considered one of the region's showplaces of multiple-use management in practice.

The Ranger himself at the time of this study, though only 35, had been in the Forest Service for fourteen years. For just two of them, however, had he been a Ranger; after his entry, he made slow progress for a decade and was once demoted after a conflict with an official from higher headquarters. But he then made deliberate efforts to improve his relations with his superiors and to get ahead; they paid off, for there was a rapid series of promotions until he was at last given charge of a district of his own; the district studied here was his first command. Somewhat belatedly, having served mostly in his native New England, he was in a new environment and in a new job trying to prove himself.

DISTRICT NUMBER TWO

By contrast, the Ranger on the second district here examined was a veteran with almost twenty-five years in the Forest Service, sixteen of them as a Ranger, the last six of the sixteen as Ranger of the district studied. This district lies in Region Eight, the Southern Region, encompassing eleven states from the East Coast south of Virginia to as far west as central Texas. The region contains twenty-six national forests (grouped into eleven administrative units) with a net area of more than 9 million acres. Like Region Seven, the land was acquired largely by purchase; much of it was marginal farm land driven out of production during the twenties and thirties, or cutover and tax-delinquent forest land, bought up by the federal government during the thirties. Old fields were planted to trees with the help of the Civilian Conservation Corps; as a result of the long growing season, many of these plantations are already yielding substantial harvests of pulpwood for the mushrooming pulp and paper industry of the South. Moreover, small sales of wood for fuel and fenceposts and other nonindustrial uses are common.

Grazing cattle and hogs on the national forests is usual in the South, particularly in the western parts of the region. Some of the land purchased by the government was occupied by tenant farmers who, for humane reasons, were not put off the holdings, and the Forest Service is thus a landlord to many of them. All this generates a large number of contacts —and problems—with local small farmers, and these contacts, along with timber management, constitute one of the major foci of attention for the members of the agency. Recreation and wildlife problems are also coming to the fore here. And fire control, under these conditions, is a challenge, especially since the trees that flourish in the region are high-hazard fuel types.

The Ranger district selected for study in Region Eight is in western South Carolina, one of three in the piedmont and mountain section of the state in a national forest not established until 1936. Although it includes only 112,000 acres of government property, it produces more timber than its Region Seven counterpart; timber management, of both saw-timber and pulpwood, and fire control are its chief functions. Most of the land was in uneconomic farms when the Forest Service acquired it and converted it to forest; stands are striking for their young and rapid growth. The example set by the Forest Service has helped transform the area; for private landowners, seeing how well the Service has done with timber, have followed its lead, and timber production is now second only to textile milling in the local economy. As a matter of fact, some of the people who sold their land to the federal government during the depression at prices then considered extremely liberal have tried, without success, to compel the government to sell it back to them at the original cost; troubles of this kind erupting now and then, and supervision of tenants still living on the district under special permits, illustrate some of the special relationships with the populace that prevail in this area.

The Ranger was hopeful of completing his career with the Forest Service on this district. After six years, he not only knew intimately the property he had to manage, but he had formed many close friendships, his wife enjoyed her teaching position in a local college, and his children had formed the especially deep attachments of the adolescent. He was doing the work he preferred in the place he wanted, and he had at that time rejected promotions rather than move; at 51, he had found the niche in life in which he would have liked to settle.

DISTRICT NUMBER THREE

The Ranger of the third district, which is in Region Nine —the North Central Region, extending over nine states from Ohio to North Dakota and down to Missouri—was not quite as content. He had entered the Forest Service almost twenty-five years earlier, rose quickly—in less than three years—to the level of Ranger, served in various places at this level (or its equivalent) for almost fifteen years, was then advanced to a higher level and assumed he was finally moving ahead, but had his hopes dashed when he was assigned again to a Ranger district after two and a half years. At the time of the study, he had been on that district for a little less than four years; since he was already 56, he had few expectations that he would ever get the advancement he once anticipated.

His district is in Michigan, one of two on a national forest that dates back to 1909, when it was created by Presidential proclamation from small, scattered blocks still in the public domain. Supplemented by purchases that began in 1926, the district now consists of about 207,000 acres (net) of government-owned land. The forest is one of fourteen in the region, most of them a mixture of reserved and (to a larger extent) purchased property comprising about 8⅓ million acres. As recently as a half-century ago, this was one of the chief lumber-producing areas of the country, thus differing from the East, where forests were obliterated in the early days of the Republic more to clear land for farms than to market the timber in them. By the time the Forest Service moved in, however, the ruthlessly exploited forest lands of this Lake States area were in neither agricultural production nor timber production; all that remained of what had once been considered an inexhaustible stand of timber were hardwood remnants; and the giant firms of the lumber industry, having exhausted their source of supply, had moved West. The Forest Service was faced with the task of restoring the natural wealth, one of the past economic mainstays, of the area.

As in Regions Seven and Eight, the task must be accomplished in the face of difficulties engendered by relationships with a heavy concentration of population; particularly in Michigan, where large industrial areas in the southern part of the state stretch northward almost to the boundaries of the national forests, this problem of resource management claims a great deal of attention. On the district chosen for study, for example, while timber planting and harvesting and fire control are the main activities, a wide variety of special permits for many purposes (including summer homes, excavation of gravel and sand, special gas and oil uses, and others) must also be handled. Recreation is growing in importance by leaps and bounds. A cluster of industrial cities—Bay City, Saginaw, and Midland—are a short distance away; Detroit and the whole industrial belt between Detroit and Chicago are only a couple of hours beyond that; and major highways to the upper Michigan peninsula pass through or close to the district bringing a larger number of visitors and travelers each year. The demands on the national forests are increasing proportionately.

DISTRICT NUMBER FOUR

Neither population nor timber is a major facet of administration on the fourth of the districts studied. For this district lies in Region Two, the Rocky Mountain region, comprehending five states (Colorado, Kansas, Nebraska, and most of South Dakota and Wyoming), fifteen national forests, and slightly under 20 million acres owned by the United States. In this section of the country, grazing of cattle and sheep is the chief use of the national forests, although water, wildlife, and recreation are also important.

The national forests of Region Two were established by Presidential withdrawal of large portions of the public domain from entry and settlement as early as 1893 and by additional reservations in subsequent years. Additional property was acquired by exchange, also, but on a small scale compared to the acquisitions east of the Mississippi; most of the exchanges were to straighten boundaries or consolidate holdings rather than to add land. By the time the reserves were created and put under management, stockmen in the area had for years been grazing their livestock on the public domain in whatever numbers they desired, and Forest Service efforts to limit the use of the range in order to put a stop to the serious damage to the vegetation cover and the consequent erosion of the soil met with stiff resistance; the Service ran head-on into local traditions. Holding grazing at safe levels without injuring the interests of the grazers is the biggest of its administrative problems. There is some timber in the region, but much of it is still rather inaccessible, and it is not likely to be fully tapped until the price of wood goes high enough to warrant the cost of constructing roads into the remote parts of the mountains. Most of the region is sparsely populated.

The district examined in Region Two is far from population centers. Lying in Colorado, just to the west of the Continental Divide (which constitutes its easternmost boundary), its headquarters is in a declining mining town of some 400 people, and the largest town in the vicinity, more than thirty-five miles away, contains only 2,500 inhabitants. Its 455,000 acres of government-owned land range from 9,000 to 14,000 feet above sea level, and are covered by snow eight to twenty feet deep from mid-fall to mid-spring; all the field work is concentrated in five months of the year.

The bulk of the field work involves range management—protecting and restoring the forage, keeping track of the herds, and maintaining relations with the stockmen. Recreation management has assumed large proportions in recent years despite the relative isolation of the district; winter sports are developing, and hunters, fishermen, and campers visit in great numbers during the other seasons. To accommodate the hunters, wildlife management is also given heavy emphasis. Administration of the water resources of the district receives high priority. Fire is not nearly so acute a problem as it is in other regions, and timber management is as yet a minor activity, although a large timber area is being opened up.

The Ranger on this district was 38 years old at the time

of the study, having entered the Forest Service seven years earlier (following five years in military service in World War II and four years in forestry school). He had become a full-fledged Ranger four years prior to the study, and had been on this district for less than two years when the research was done. Born and bred not far from the Ranger station he occupied, he had wanted from his youth to become a Forest Ranger; his rapid advancement to this position gave him great pleasure and built up his hopes for his future in the agency.

DISTRICT NUMBER FIVE

His experience was somewhat similar to that of the Ranger on the fifth of the districts studied here. This man decided on a career in forestry comparatively late in life; he loved outdoor work, but did not become aware of the profession until he had served four years in the Army Air Force in World War II and worked four more years as a journeyman carpenter. But when he discovered forestry, he pursued this career enthusiastically. At the age of 26, he entered forestry school, completed his undergraduate work at 30, entered the Forest Service and simultaneously completed a year of graduate work. A year later (that is, two years after joining the Forest Service), he was made a Ranger, and had been at that level for almost three years at the time of the study; less than a year of this time, however, was on the given district.

His district is in Region Six, the Pacific Northwest Region, consisting of nineteen national forests with a net acreage of over 23 million, mostly withdrawn by Presidential proclamation from the public domain, in Oregon and Washington. More than 36 per cent of all the saw timber in the country stands in this region, and more than 41 per cent of the saw timber in the region is in the national forests; much of it is still virgin forest. As saw timber plays out on private lands, the demands that the Forest Service sell more of its timber than it considers safe have grown more clamorous, and pressures directed at the regional office in Portland, at the Washington office, the Secretary of Agriculture, and Congress have been heavy and persistent. While there is considerable grazing on the national forests in the region, managing timber, and protecting it against fire and disease, are the chief activities here, and with them have come the problems of relationships with many of the largest lumber companies and trade associations in the country. In addition, visits by recreationists have increased many times over since World War II.

This district (which has a net, government-owned area of 131,000 acres) is an exceptionally interesting one silviculturally because it is a zone of three kinds of transitions: located in southern Oregon, it is an area in which the species of northern California give way to the species of the Northwest; lying on the upper, western slope of the Cascade Mountains, it is an area in which the species of the eastern slope peter out and those characteristic of the western slope begin to take over; being between 2,200 and 6,800 feet in altitude, it is an area in which the coastal, rain-forest species yield to species that flourish at greater heights. Altogether, it harbors twelve kinds of merchantable trees, including giant Douglas fir, Ponderosa pine, Sugar pine, Western hemlock, California cedar, and Shasta red fir—trees so large that the tops disposed of as slash are often equal in size to trees harvested east of the Mississippi. Timber management and fire control are therefore the dominant functions on the district, with recreation and grazing following not far behind.

ADEQUACY OF THE SAMPLE

The five districts thus present a considerable spread of conditions under which administration of the national forests is conducted, and the Rangers themselves constitute a variegated sample of field officers. While the selection is admittedly far from exhaustive of all the possibilities, it seems sufficiently diversified to justify the assumption that the influences on administrative behavior common to all of them are not "sports" (in the biological sense of sudden, spontaneous variations from type); the common features, intuitively speaking, would appear to have wider applicability, to suggest the major patterns of influence in the Forest Service. It is with these patterns that the study is concerned.

Method of Study

Prior to the field research, each Ranger was asked to send specified materials about himself and his district, including biographical data, employment histories, workload and financial statistics, tables of organization, inspection reports, work plans, characteristics of his locality, and other information. This, coupled with the general background reading and research and some preliminary interviews in Washington and a Ranger district close to research headquarters, made it possible to concentrate field work on the specific problem on which the study is focused. The advance preparation provided considerable familiarity with field conditions.

Then, each district was visited—the first for several weeks in order to establish a pattern, the others for a week each. The Rangers put themselves fully at the disposal of the researcher during his visits; all their time was set aside for the discussions. In addition, social visits in the evenings offered opportunities to continue the discussions not only with the Rangers but with members of their families as well, under quite informal circumstances.

"Conversations" would be as accurate a way of characterizing the discussions between the researcher and the Rangers as "interviews." Although the talks were eventually brought around to all the subjects taken up later in this volume, they were not forced into any preconceived mold. Nor were they simply sessions of questions and answers; ideas and opinions were exchanged, often vigorously, and inconsistencies between practices observed or reported on one district and those found or described on others were explicitly pointed out for explanation.

Washington office permission for the study greatly facilitated this process. Although the Rangers knew Washington did not sponsor, direct, or control the study, and had not commissioned it or requested it, but had only as a favor allowed an academic researcher to conduct it, the fact that the researcher had obtained prior approval from Washington, and that Washington had authorized the Rangers to give the project all the time required for it, indicated to them they were not acting indiscreetly by talking at length about internal agency affairs with an outsider. Moreover, pains were taken to let each Ranger know the manuscript would be discussed with Washington officials before it was released, so the field men did not feel they were carrying the full responsibility for safeguarding the interests of the Forest Service. And, at the same time, each Ranger was assured he would have an opportunity to review relevant portions of the manuscript before it was shown to any of his superiors, so that no Ranger would do himself any professional injury by talking freely;[6] this pledge also served to dissociate the researcher from the Washington office, and to identify him as a disinterested academic rather than as an investigator of some kind.

Despite these precautions, several of the Rangers were wary; they opened up only when it became apparent to them

[6] Further to protect the Rangers, a measure of ambiguity about the individual sources of particular statements has been deliberately maintained throughout the text of the chapters that follow, although the sources are identified in the author's files.

that the researcher was sufficiently familiar with the organization to detect reticency and that they could use the familiar jargon and abbreviations and make many of the same assumptions they did when they were talking with one another. Before each week-long session was over, most of the Rangers raised questions and expressed complaints they confess they would have hesitated to voice to anyone inside the Service. The atmosphere of the conversations was generally cordial and confidential.

The conversations were not confined to the Rangers and their families. In every case but one, there were discussions with forest supervisors (chief officers of individual national forests, and the immediate superiors of the Rangers) and members of the supervisors' staffs. In one instance, it was possible to hold long, detailed, informal talks with a former Ranger who had been away to study public administration and returned to duties as a management analyst. On every district, the Rangers' aides and assistants were consulted, and, on a few, there were opportunities to speak at some length with users of national forest products and services.

The discussions were supplemented by observation as often as possible. Tours were made through the districts to see the physical properties under Ranger jurisdiction and to study in action some of the operations that make up resource management. The researcher sat in on some conferences between Rangers and their staffs, Rangers and supervisors, Rangers and the public. Files and records were thrown open to examination.

On the theory that note-taking inhibits free discussion, it was kept to a minimum. Instead, at the end of each day, the material gathered that day was recorded on tape and later transcribed.

Finally, the field notes, the background reading, and the data collected on visits to the Washington office were assembled and analyzed. In the end, a great many of the findings remained highly impressionistic. But the impressions in this study are based not merely on library sources or the word of agency leaders; whatever their shortcomings, they are rooted in relatively intensive examinations of men and conditions in the field. And whatever their shortcomings, they bring into sharp relief some aspects of Forest Service administration not often remarked by other commentators.

The Plan of the Book

What follows is a composite picture of the Forest Service. Part One describes the tendencies toward fragmentation of the Forest Service, the factors that operate to break it up into its component units and render each an independent miniature of the agency as a whole. Chapter II takes up the centrifugal tendencies inherent in the magnitude and character of the work of the Forest Service, and in the characteristics of the organization required to discharge its responsibilities. Chapter III deals with centrifugal tendencies with which most large organizations, regardless of the nature of their work, must contend, and with some special elements in Forest Service history that accentuate these tendencies.

Part Two then lists the factors that tend to hold the Forest Service together, to integrate the activities of its personnel, to forge a national policy in the face of circumstances conducive to many local, unrelated policies. Chapter IV treats the means by which field decisions are in effect made in advance for field officers. Chapter V discusses the methods of holding field behavior in the channels thus established regardless of the preferences of field officials. Chapter VI examines the techniques that result in field officials doing of their own volition the things the leaders of the Forest Service want them to do.

Ticking off variables in this fashion facilitates presentation and analysis, but it does not portray the interrelationships of the variables; loosely speaking, it suggests a series of still pictures quite inadequate to depict a dynamic organization in action. Hence, Chapter VII is an attempt to put the pieces together, to capture them in motion, to study their interplay, and, from this analysis, to derive the lessons to be learned from Forest Service experience and from this way of approaching it.

Three things, it is hoped, will emerge. One is a description of the administration of the national forests as an individual operation that gives some fresh insights into the behavioral dynamics of a specific large-scale organization. The second is a set of generalizations about organizations in general, propositions that can be tested against the experience of other organizations of all kinds. The third is a series of conclusions about the utility of looking at organizations from the viewpoint here proposed and attempted.

PART ONE

TENDENCIES TOWARD FRAGMENTATION

CHAPTER II

THE SIZE AND COMPLEXITY OF THE FOREST SERVICE JOB

A Profile of the Forest Service

Even in agencies with simple, routine responsibilities, welding the behaviors of field personnel into integral patterns is often a trying experience. If, in addition, the field personnel operate under widely varied conditions, the difficulties of integration are multiplied many times. If, furthermore, the field units are too scattered to readily permit close supervision of their activities by agency leaders or co-ordination by personal contact among the men in the field, the difficulties increase exponentially. And if their responsibilities have not grown gradually, over many generations, enabling the members of the organization to work out their adjustments slowly, the administrative burden may be staggering.

All of these hardships beset the Forest Service. Its tasks are complex and require the exercise of broad discretion. No two of its field units face precisely the same problems, and conditions in some parts of the country barely resemble those in others. The tracts managed by the Service are scattered across the breadth of a continent and beyond. Its assignments have been thrust upon it in a relatively short space of time. Great indeed are the challenges to its administrative officers.

DEVELOPMENT AND PRESENT SCOPE OF RESPONSIBILITIES [1]

Not until 1876—a full century after the signing of the Declaration of Independence—was there even the germ of an administrative agency for forest management in the United States government, and even the germ planted in that year was not especially robust. A few statutes enacted prior to this had the effect of protecting timber on federal lands from depredations, and one encouraged the planting of trees on western prairies, but no administrative machinery was established. Indeed, forests were long regarded merely as impediments to the westward flow of population; clearing them away was more important than protecting them. Besides, the "legend of inexhaustibility," according to which the forests were too extensive to be depleted by the feeble efforts of mankind, held sway. So it was a hundred years before a forestry organization was set up.

The beginnings were modest indeed—a special unit established by administrative action in the Department of Agriculture, with appropriations of just $2,000, to gather historical and statistical information about forests and forest products. In 1881, however, it was elevated to the status of an administrative division, and in 1886 received permanent statutory rank. Appropriations increased over the years, but they were still only $48,000 at the turn of the century. Under Bernhard E. Fernow, the Forestry Division built a name,

[1] This historical summary is based on J. Cameron, *The Development of Governmental Forestry in the United States* (Baltimore: The Johns Hopkins Press, 1928); J. Ise, *The United States Forest Policy* (New Haven: Yale University Press, 1920); D. H. Smith, *The Forest Service* (Washington: The Brookings Institution, 1930). See also G. Pinchot, *Breaking New Ground* (New York: Harcourt, Brace and Co., 1947); U. S. Forest Service, *Highlights in the History of Forest Conservation* (U. S. Government Printing Office, Agriculture Information Bulletin No. 83, 1952). A "short popular account of the work of the United States Forest Service on the National Forests" is R. H. D. Boerker, *Our National Forests* (New York: The Macmillan Co., 1918).

expanded its research activities to include technical forestry subjects as well as history and statistics, and undertook a vigorous campaign of public education. Nevertheless, it still had no part in actual operations in the woods.

Fernow resigned in 1898 and was replaced by Gifford Pinchot. In three years, the division was reorganized and became the Bureau of Forestry, furnishing a vehicle for Pinchot to implement his philosophy of the role of government in relation to forestry, a philosophy that shaped and colored the relationship for years to come and also left a permanent impress on the organization. Research and information were all right as far as they went, but the real work of a public forestry agency, as he saw it, had to be centered on work that went on in the field. He therefore initiated a program of assistance to private owners of forest lands and started the training of personnel in practical forest administration; by 1905, the appropriation of the Bureau had risen to more than $439,000. In that year, the young forestry agency and its energetic leader were given forests to manage; the Forest Service was born.

Up to 1905, jurisdiction over federally owned forests was lodged in the Department of the Interior. Initially, the jurisdiction was derived from legislation dating from the early nineteenth century providing for protection of forests on the public domain from theft. In 1891, however, a landmark law was enacted authorizing the President to set aside portions of the public domain as forest reserves; these lands were not open to entry or settlement, and it was the Interior Department that guarded them. The authority of the Secretary was broadened by an historic statute that became law in 1897 and furnished the foundation for modern forestry practices in the federal government; it authorized him to control and administer the reserves as well as to protect them, and conferred on him the power to make rules governing their occupancy, use, and disposition. These reserves and these powers passed to the Department of Agriculture in 1905.

The transfer, largely the result of skillful efforts by the man who was to become the first Chief—Gifford Pinchot—put an end (over strenuous Interior Department objections) to the anomalous division of functions that existed between 1876 and 1905. On the one hand, the specialized forestry agency in the Department of Agriculture had charge of forestry but not of forests. On the other hand, the Department of the Interior was responsible for the public forests but not for developing forestry. The merger of these functions in one department put an end to this strange duality: the Bureau of Forestry took over the task of protection and management in addition to its programs of research, assistance, and public information; the expanded organization was officially redesignated the Forest Service; and the reserves were legally named "National Forests."

The area of the national forests was rapidly increased by executive withdrawals of land from the public domain under the act of 1891. Most of these additional reserves were created in the West, where the greater part of the property to which the federal government still held title lay. By 1911, however, attention turned to the East, where whole areas had been completely stripped of trees by loggers, or burned over to make room for farms which were abandoned as better farm land became available farther west and drove marginal farms out of production. It was recognition of the flood control problems resulting in part from these conditions that led to the enactment of the Weeks Act in 1911, which set up the National Forest Reservation Commission and authorized the Secretary of Agriculture to purchase land for addition to the national forest system, provided that such purchases were approved by the Commission and by states in which they were made.

The Forest Service was thus able to move into the East, and the last major legal gap in the powers needed for administration was closed. "1876, 1891, 1897, 1905—these were all red-letter forest years. But 1911, a year of the same hue, deserves somewhat larger type, for it puts the seal of finality upon them all." [2] The legislative base was complete; no other additional pieces of legislation would or could add such large increments to the jurisdiction of the Forest Service, although important measures did round out its authority in later years.

Developing the maximum practical yield and beneficial use of the national forests has from the start been an awesome responsibility. It began in 1905 with 60 million acres concentrated in the West, and the area of forest land under Forest Service jurisdiction has since then been increased to 181 million acres spread over 42 states and a territory. In addition, the Forest Service has been made responsible for administering about 7 million acres of land utilization projects (designed to restore the soil in areas once in submarginal farms which were purchased by the federal government and retired from agriculture during the Great Depression). National forest administration at once became the largest of the Forest Service programs, a position it has never lost.

But the old programs were not displaced; on the contrary, they took on added vigor.[3] For it quickly became apparent that the national forests could not solve the nation's forest problems because they encompassed only part of all the commercial forest land in the United States; the vast bulk of the commercial forest land was privately owned, and some was in the hands of state and local governments; so the practice of co-operation with, and assistance to, private owners and states and localities, instituted before the turn of the century, not only had to be continued, but eventually intensified. Likewise, the research programs begun in the nineteenth century became increasingly important; both proper management and

[2] Cameron, *op. cit.*, p. 284.
[3] See section below.

effective advising and co-operation demanded more information about the technical and economic aspects of forestry than had ever before been required. So the activities that preceded national forest administration expanded and flourished under the impetus of the newer function. To this day, the program of the Forest Service remains three-pronged.

VOLUME AND VARIETY OF ACTIVITIES[4]

Carrying out this three-pronged program is, as of 1956, a $128-million operation (not counting an additional $28 million distributed to states, territories, and counties, mainly for roads and schools). At the same time, the Forest Service takes in (during 1956) over $137 million in receipts, from three sources. First and foremost, there are receipts from the sale of forest products (chiefly timber) and from fees for permits for grazing and other uses of the national forests ($118 million in 1956). Second, funds (over $11 million in 1956) are deposited with the Forest Service by private organizations and individuals—by timber operators and other national forest users to pay for improvement of timber stands or ranges, road construction, brush disposal, and other work that the users would otherwise have to do themselves, meeting Forest Service specifications; by community organizations and owners of private lands for a variety of services to communities and landowners not directly connected with the national forests, such as production and distribution of planting stock, technical forestry assistance, and some kinds

[4] This section draws mainly on the *Report of the Chief of the Forest Service, 1956,* and earlier reports in the same series; U. S. Forest Service, *Timber Resource Review* (preliminary review draft, processed, 1955), Chapter IX; U. S. Congress, 84th Congress, 2nd Session, House of Representatives, Committee on Appropriations, Subcommittee on Interior and Related Agencies, *Hearings*, 1956, pp. 555 ff.; U. S. Congress, 84th Congress, 2nd Session, Senate, Committee on Appropriations, Subcommittee on Interior Department and Related Agencies, *Hearings*, 1956, pp. 571 ff.

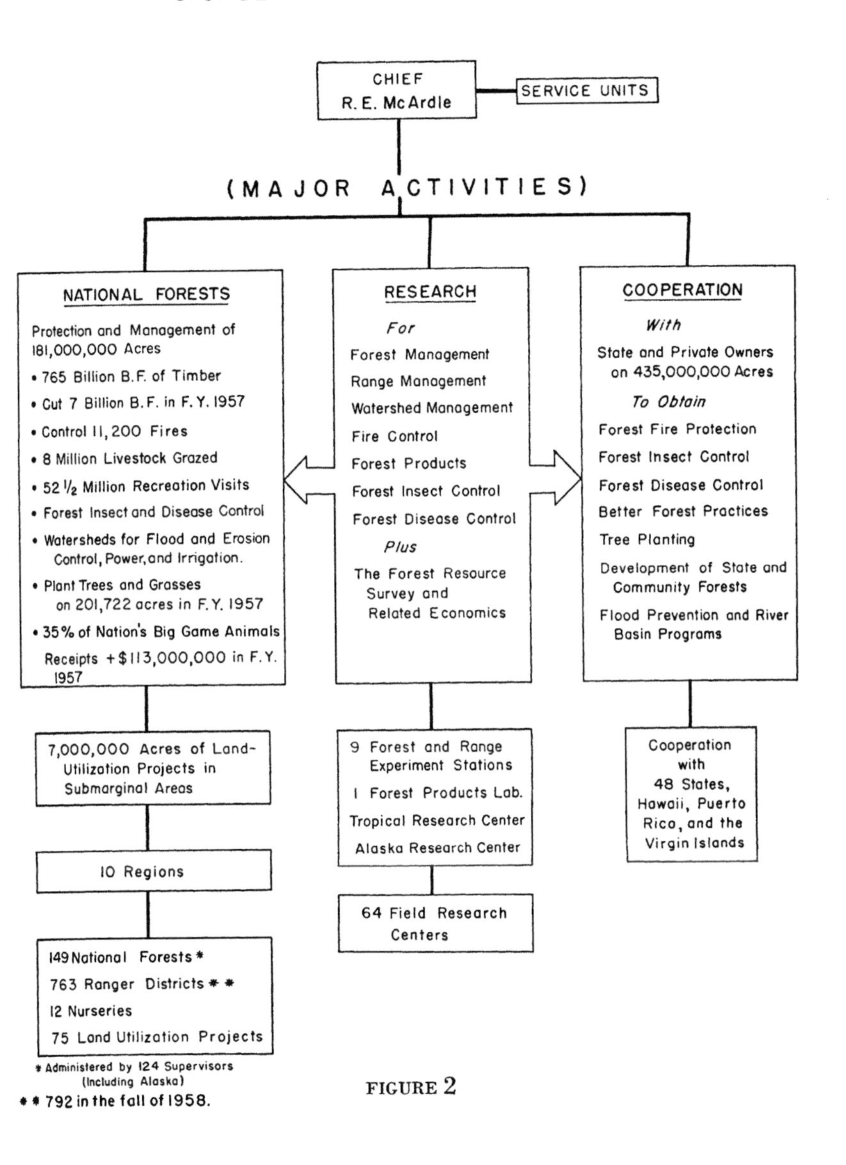
FOREST SERVICE
U. S. DEPARTMENT OF AGRICULTURE
CHIEF
R. E. McArdle
SERVICE UNITS
(MAJOR ACTIVITIES)
NATIONAL FORESTS
Protection and Management of 181,000,000 Acres
• 765 Billion B.F. of Timber
• Cut 7 Billion B.F. in F.Y. 1957
• Control 11,200 Fires
• 8 Million Livestock Grazed
• 52 1/2 Million Recreation Visits
• Forest Insect and Disease Control
• Watersheds for Flood and Erosion Control, Power, and Irrigation.
• Plant Trees and Grasses on 201,722 acres in F.Y. 1957
• 35% of Nation's Big Game Animals
Receipts + $113,000,000 in F.Y. 1957
RESEARCH
For
Forest Management
Range Management
Watershed Management
Fire Control
Forest Products
Forest Insect Control
Forest Disease Control
Plus
The Forest Resource Survey and Related Economics
COOPERATION
With
State and Private Owners on 435,000,000 Acres
To Obtain
Forest Fire Protection
Forest Insect Control
Forest Disease Control
Better Forest Practices
Tree Planting
Development of State and Community Forests
Flood Prevention and River Basin Programs
7,000,000 Acres of Land-Utilization Projects in Submarginal Areas
10 Regions
149 National Forests *
763 Ranger Districts * *
12 Nurseries
75 Land Utilization Projects
9 Forest and Range Experiment Stations
1 Forest Products Lab.
Tropical Research Center
Alaska Research Center
64 Field Research Centers
Cooperation with 48 States, Hawaii, Puerto Rico, and the Virgin Islands
* Administered by 124 Supervisors (Including Alaska)
* * 792 in the fall of 1958.

FIGURE 2

of research. Third, funds come from other public agencies to reimburse the Forest Service for work it has done for them.

The receipts, however, are not freely at the disposal of the Forest Service. Under federal statutes, a portion of them is earmarked for distribution to states, territories, and counties as noted above.[5] The deposited funds, under authorizing legislation, are held by the Forest Service to be spent for the purposes for which they were deposited. The rest is returned to the U. S. Treasury, whence, like all other public moneys, it becomes available for expenditure by government agencies only upon appropriation. Thus, the Forest Service, like most other federal bureaus, depends primarily upon appropriations for the bulk of its funds.

Almost three-quarters of the $128 million (roughly, $95 million) spent in 1956 on national forest administration, state and private forestry, and research is applied to the immediate protection and management of the national forests (including road construction and maintenance, and relatively small sums for acquisition of additional land or for adjusting

[5] "Under the forest reserve fund act of May 23, 1908, and amendments, 25 per cent of the national-forest receipts deposited to that fund, excepting some special allocations, is returned to the States and is allotted by the States to counties or parishes on the basis of national-forest acreage, for expenditure as provided under State law on schools and roads. A total of $27,893,210 was returned to the States under this provision; $129,404 was paid to Arizona and New Mexico school funds under the act of June 20, 1910; $46,497 was paid to the State of Minnesota under the act of June 22, 1948. Counties were paid $459,795 under title 3 of the Bankhead-Jones Farm Tenant Land Act; and $142,046 was paid to Alaska, from fiscal year 1956 Tongass Indian land receipts under the Act of July 24, 1956. In addition $455,087 was paid to Alaska under this Act representing 25 per cent of receipts collected for fiscal years 1948 through 1955, formerly held in escrow.

"Of the $2,485,782 receipts collected from national-forest revested Oregon and California Railroad grant lands . . . 75 per cent was returned to the counties concerned, as provided for under the act of June 24, 1954.

"An additional 10 per cent of the base receipts, $11,398,645 was appropriated to the Forest Service for expenditure on roads and trails within the national forests." *Report of the Chief of the Forest Service, 1956,* p. 19.

boundaries). Various kinds of co-operation with state and local governments (chiefly, for fire protection, but for other aspects of forestry as well) and with private organizations and individuals (to help them improve their forestry practices, both in economic and silvicultural terms) account for a little under 12 per cent additional. Research makes up about 7 per cent of the total. Thus, almost 94 per cent of all the expenditures goes for these three big programs. The remainder is spent on closely related activities, such as flood control, brush disposal, blister rust control, and pest control, performed to help other agencies (on a reimbursable basis).

Even the smallest of the three main programs (in terms of expenditures)—research—illustrates the widely varied nature of Forest Service work. It comprises forest-management research aimed at improving forest production by basic inquiries into the growth of trees and forests and by testing the application of these findings to the operation of forest properties. Thus, for example, projects have been conducted to determine desirable cutting methods by studies of seed dissemination, to reduce timber losses caused by storms, to improve the quality of timber by experimenting with tree breeding and selection and location, and to convert low-quality hardwood stands to pine. The research program also includes investigations of systems of forest fire control, including statistical reports and studies of fire damage. Another extensive series of projects is designed to discover the interrelationships of soil, plants, and water, and "includes the design and testing of improved cutting, logging, grazing, roadbuilding, and other practices to reduce harmful erosion, flood flows, and debris movements, and to increase the yield and quality of water supplies." Range research is carried on in order to find ways of converting low-value brush fields to grass, to find appropriate balances between forage for game as well as for livestock, and to better the vegetation for sheep that graze salt-desert shrub ranges.

Important studies of forest economics are conducted by the Forest Service, covering, among other things, timber supply and marketing and industries, such as pulp milling, that depend on forests for their raw materials. Research in forest products has developed new products, lowered costs, and increased the serviceability of existing products, reduced unused residues and found useful outlets for unavoidable residues, and solved other forest products problems; this phase of research encompasses such projects as tests of pulping behavior of different woods, experiments on the strength of wood at low temperatures, accumulation of information on preservation of wood in glued products, and studies of fire hazards in houses. Thus has the original function of the old Forestry Division of the Department of Agriculture burgeoned over the decades. Yet comprehensive and diversified as it is, it constitutes only a fraction of the activities under the jurisdiction of the leaders of the Forest Service today.

Co-operative programs make up a larger portion. For of the more than 664 million acres of forest land in the United States—covering one-third of the total land area of the country—488 million acres now bear, or are capable of bearing, merchantable timber, and of these 488 million acres of commercial forest land, only a little over 17 per cent are in the national forests. Private owners hold more than 73 per cent of it, and state and local governments own almost 6 per cent; federal agencies other than the Forest Service have the remaining 4 per cent. Consequently, it is clear that a bureau committed, as the Forest Service is, to maximizing the use and protection of the nation's forest resources must reach out to influence the management and protective practices of private owners, of states and localities, and, at times, of other federal agencies. Since there are in excess of 4.5 million private owners of commercial forest land, the vast majority of whom possess under 100 acres, but whose holdings in some cases are larger than 50 thousand acres, this is obviously no small task. It is one of the responsibilities of the Forest Service.

This responsibility is discharged by means of a number of co-operative programs. In most of these, the Forest Service acts as a partner to co-operating state agencies; the state agencies employ the personnel who actually conduct the programs, and manage the programs in accordance with their established procedures, while the Forest Service allots the federal funds that cover part of the expenses, and serves as an adviser with respect to the disposition of the federal share of the expenditures. For example, under recent legislation, the Forest Service apportions federal funds among 38 state forestry departments that co-operate with it "in providing on-the-ground technical assistance to owners of private forests and to small sawmill operators and other processors of primary forest products." It also works with the Extension Service of the Department of Agriculture, the land-grant colleges, and state extension services to reach farmers with woodlots on their land, and it co-operates with the states in distributing forest-planting stock in order to encourage and facilitate afforestation and reforestation for flood, water, and erosion control as well as for timber production. Under the provisions of the Conservation Reserve aspects of the Soil Bank, the Forest Service, through the state forestry organizations, furnishes funds for production of forest tree planting stock and for technical assistance to landowners.

Further, it administers a program of aid to gum turpentine farmers who follow acceptable conservation practices as part of a naval stores conservation program. It co-operates also with most of the states in programs of fire suppression, for which it distributes federal funds, and it has joined hands with the Advertising Council, Inc., in an information and education crusade for fire prevention. In forest pest control and in flood prevention, the Forest Service works with state agencies and departments. In one program—assistance to owners of large forests, whether private or governmental, a type of service not usually available from the states—it deals directly with individuals without going through the states, but it customarily operates in close contact with state officers in order to demonstrate its methods and to signalize the desirability of eventual state entry into this field of service. In the same way as forest administration and research, the Forest Service programs of co-operation and assistance and education are many-faceted, intricate in composition, and spread across the whole country.

But administration of the national forests towers over the other programs in fiscal size, administrative complexity, and geographic spread. Protecting and managing them is the giant among the activities of the Forest Service.

Because they are called national *forests*, it is natural to think of the great tracts managed by the Forest Service in terms of timber. Indeed, since an estimated 765 billion board-feet of timber stand on them, from which 7 billion board-feet, worth $107 million, were cut in 1956, this is certainly not a misleading inference. But it would be equally appropriate to refer to the national forests as national pastures, national playgrounds, national water reservoirs, and national wildlife centers, for 8 million head of livestock grazed on these lands in 1956, under grazing permits issued by the Service. An estimated 53 million recreational visits were made to the national forests, including millions for hunting and fishing, for which the Service provides many facilities. Waters rising in the national forests are immensely important not only to the forests themselves but to areas far beyond their boundaries.[6] Over one-third (about 2 million)

[6] "The water that goes over Grand Coulee Dam, that goes into Boulder Lake, that goes into every big reservoir, whether it is for power or reclamation or flood control, in the western parts of the country, originates principally on the national forests. . . . I think every city in the West . . . gets its domestic water from streams originating in the national forests or from . . . the national forest watersheds. So that the whole

of the big-game animals in the United States, as well as thousands of fur bearers, small-game animals, game birds, and waterfowl frequent the national forests, not merely by chance, but, in considerable part, because the Forest Service manages their habitats so as to encourage their increase without damage to the other uses of the lands. And there are also summer homes, summer and winter sports and resort areas, mines, sand and gravel pits, quarries, reservoirs, and other installations under permit in the national forests. All this, taken together, is what is meant by "multiple-use management," one of the foundation stones on which Forest Service policy is built. And this is why national forest protection and management loom so large in its threefold program.

Performing this task requires making and supervising 30,000 timber contracts a year, as well as issuing and checking on more than 17,000 grazing permits and 56,000 special-use permits covering some 110 types of uses. It necessitates controlling as many as 12,000 fires and planting as much as 200,000 acres in trees and grasses a year. It calls for the maintenance of 75,000 miles of roads and 115,000 miles of trails in and through the national forests, and the construction of several hundred additional miles of road and miles of trail, as well as over 500 bridges, each year. It makes necessary the building and maintenance of camping sites, picnic areas, fireplaces, and safe drinking water facilities. It involves prevention and punishment of trespass. Concrete actions such as these add up to the sum and substance of Forest Service policy; everything else is only intention. (And every one of these actions, every one of these facilities, occurs inside the boundaries of a Ranger district. Whatever is done,

western water fabric is conditioned on flows from the national forests... (We) have found . . . that there are considerable possibilities in the manipulation of our timber cutting practices, so as to increase the flow of water from certain types of watersheds by as much as a third." U. S. Congress, 80th Congress, First Session, Committee on Appropriations, Subcommittee on the Department of Agriculture, *Hearings*, 1947, p. 53.

where, to whom and for whom, how fast, by what method—the critical decisions in policy formulation—it is executed and also influenced by the Rangers in the field.)

The observation that these programs are spread over 181 million acres of land scattered through 42 states and a territory only suggests the variegated conditions under which the Rangers function. Uniform conditions would make it comparatively easy to integrate Ranger activities—still difficult under any circumstances, yet easier, relatively, than it is now. But the national forests

> range from the pines of the deep South and the southern hardwoods of the Appalachians to the spruces and pines of the White Mountains in New England and the northern hardwoods of the Lake States; from the piñon and juniper stands in the southern Rockies of New Mexico to the pine and fir forests along the Canadian line in Montana and Idaho; from the chaparral-covered foothills of southern California to the great conifer stands of the Olympic and Cascade Mountains in northern Washington. In Alaska, . . . valuable Sitka spruce and hemlock clothe the lower flanks of the coastal mountains . . . In Puerto Rico, . . . great trees hung with vines and lianas spread their immense crowns over a steaming tropical jungle of brilliant flowers and moss.[7]

Into some of these areas pour the millions of recreation-seekers, while other areas, many of them reaching above timber line, are rarely visited at all. In some, timber production is the principal activity; in others, range management dominates. A few forests are still virgin; others have been cut and burned over for centuries. In many areas, the Forest Service is welcomed, and wood-using industries and private owners of forest lands co-operate enthusiastically with it; in others, powerful interests fight to get control of the resources it has been managing in order to intensify their

[7] U. S. Forest Service, *The Work of the U. S. Forest Service* (U. S. Government Printing Office, Department of Agriculture Miscellaneous Publication No. 290, Revised 1945), p. 5.

exploitation. In densely populated places, manpower for fire fighting and other temporary and seasonal jobs is easy to come by; in others rounding up crews can be a major problem.

Communications via commercial facilities are simple in some forests; the Forest Service must string its own lines and set up its own radio links in others. Working out relationships with state and local governments, and with local populations, places a heavy demand on the time of Service personnel in some parts of the country; this is a negligible burden in others. Climate varies. Vegetative types differ. There is a great spread in fire risks and hazards from place to place. Different industries have different needs and make different demands (e. g., pulp and paper factories in some areas, sawmills and lumber producers in others). Variations in local economic situations cannot be ignored. The list could be extended almost indefinitely, but there is no need to labor the argument; beyond any question, no two areas are identical, and many hardly resemble others. This adds to the complexity of National Forest protection and management, which size and multiple-use objectives alone would have made difficult even without such variety. The third and largest prong of the Forest Service program thus presents the greatest administrative challenges of all. Fusing the behaviors of field personnel in national forest administration into a coherent pattern requires the conquest of enormous obstacles.

ORGANIZING TO PERFORM THE TASKS

Formally presiding over the organization developed to discharge (and to define) the responsibilities of the Service, is the Chief. The Secretary of Agriculture appoints the Chief of the Forest Service, but, the position being in the competitive classified service, appointments are reviewed by the Civil Service Commission, which has the authority to disapprove candidates who do not meet the established standards for the office. In practice, it is normally filled by promotion of an assistant chief.[8] Almost all chiefs have served until retirement or death.[9]

The Chief constitutes the apex of a traditional administrative pyramid—or, perhaps more accurately, what amounts to two pyramids. One is organized for the administration of the national forests and co-operation with the states, local governments, and private forest owners; it consists of ten regional foresters, 124 forest supervisors (who manage a total of 149 national forests, some of which are combined for administrative purposes), and 792 district Rangers. The other is organized for research, and is made up of nine directors of experiment stations, 51 research centers under them, 102 experimental forests and 14 experimental ranges, and three special units (the Forest Products Laboratory, the Tropical Forest Research Center, and the Alaska Forest Research Center) reporting directly to Washington. All three major programs of the Service are handled by these two hierarchies.

Six assistant chiefs aid the Chief. One concentrates on National Forest Resource Management, another specializes in Research, and a third is assigned to State and Private Forestry; there are assistant chiefs for Administration (who handle organization and methods projects, budget and finance, personnel administration, inspections, administrative services, and public relations), for Lands (i.e., land acquisition and uses), and for Program Planning and Legislation. Also attached directly to the Chief's office is an Internal Audit Division, which has service-wide responsibility for all Forest Service audits (including appraisal of policies, plans,

[8] There have been a couple of exceptions; see A. W. Macmahon and J. D. Millett, *Federal Administrators* (New York: Columbia University Press, 1939), pp. 346-350.

[9] Only one Chief has ever been removed. Gifford Pinchot, the first head of the Service, was dismissed by President Taft in 1911 as a result of his controversy with Secretary Ballinger of the Department of the Interior.

FOREST SERVICE ORGANIZATION LINE OFFICERS

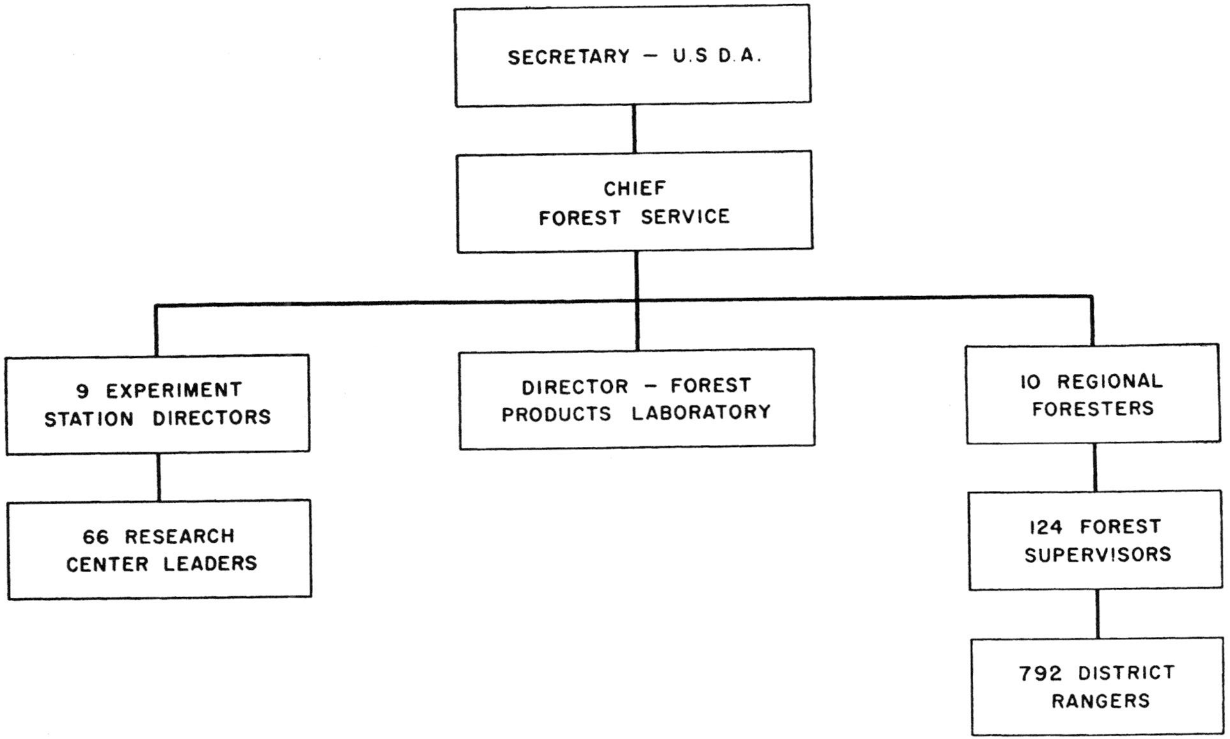

FIGURE 3

ORGANIZATION CHART
FOREST SERVICE
U. S. DEPARTMENT OF AGRICULTURE

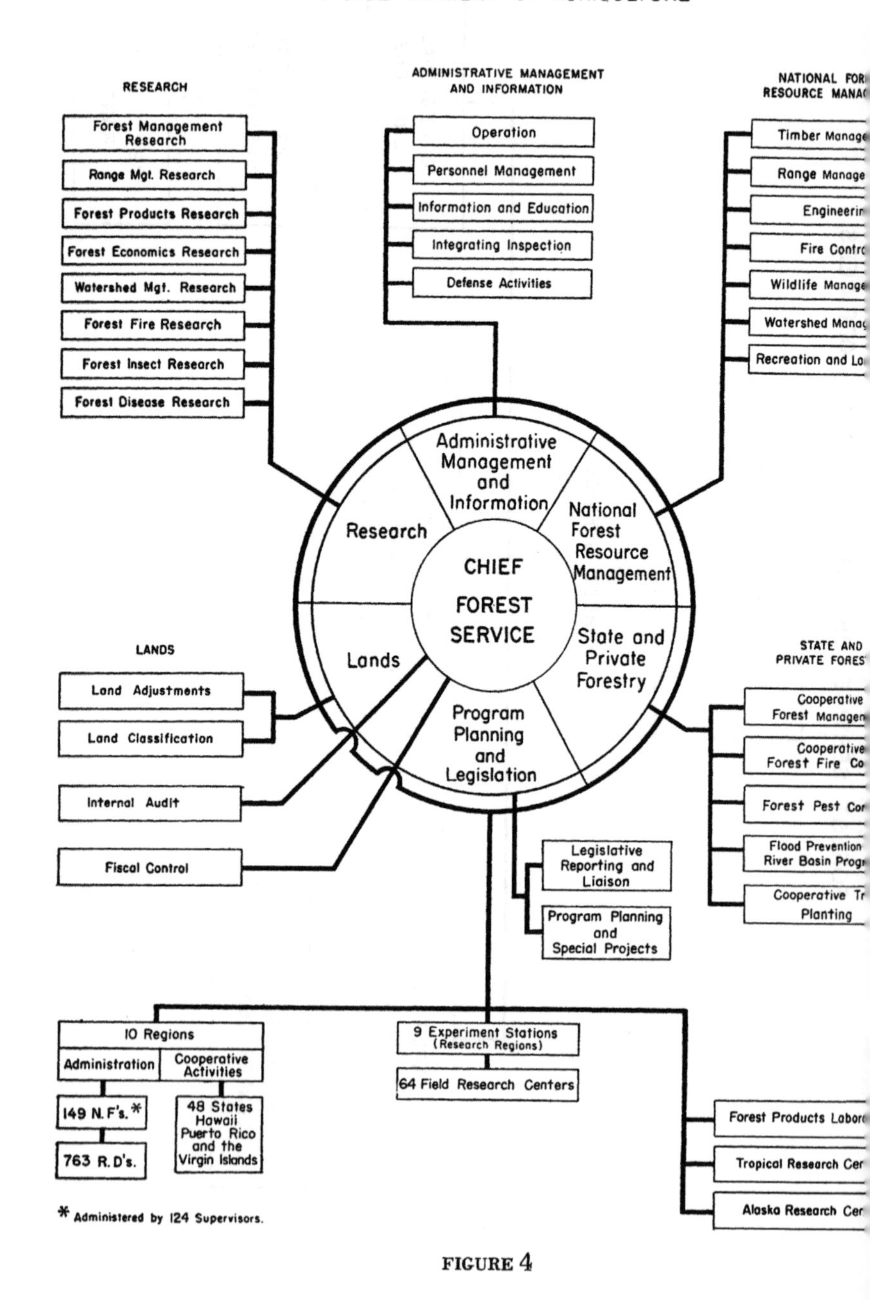

FIGURE 4

procedures, and efficiency).

The Chief and his six assistant chiefs are commonly treated as a single administrative unit, referred to as "staff." Each month, a different assistant chief assumes the responsibility for serving as acting chief. Theoretically, the assistant chiefs are merely arms of the Chief, extensions of his official will and personality, and everything that emanates from staff is to be regarded as coming from the Chief. (In practice, it will be seen later on, things are not quite this simple and clear-cut.)

The six assistant chiefs head highly specialized staffs. The Assistant Chief for Research has eight divisions under him; the one for National Forest Resource Management has seven; Administration, five; State and Private Forestry, four; Program Planning and Legislation, two; and Lands, two. The divisions are not engaged in field operations; they do not do the physical work of the agency. Rather, they are supposed to be compilers of information, sources of ideas, and observers of field work. They are neither on the firing line nor in command.

Below the Washington level are the directors of the experiment stations and the regional foresters. The work of the research branch of the administrative family, though important, is not germane to the discussion of national forest administration, and will not be explored further. The regional foresters, however, are immediately relevant to this study, for they command the field forces in the national forests.

Each of the regional foresters is responsible for all the functions of the Forest Service, except research, within his own region. At the regional level, these are defined in the agency as: timber management, fire control, range and wildlife management, recreation and lands uses (including land acquisition and watershed management), engineering (construction planning and execution), operation (administrative

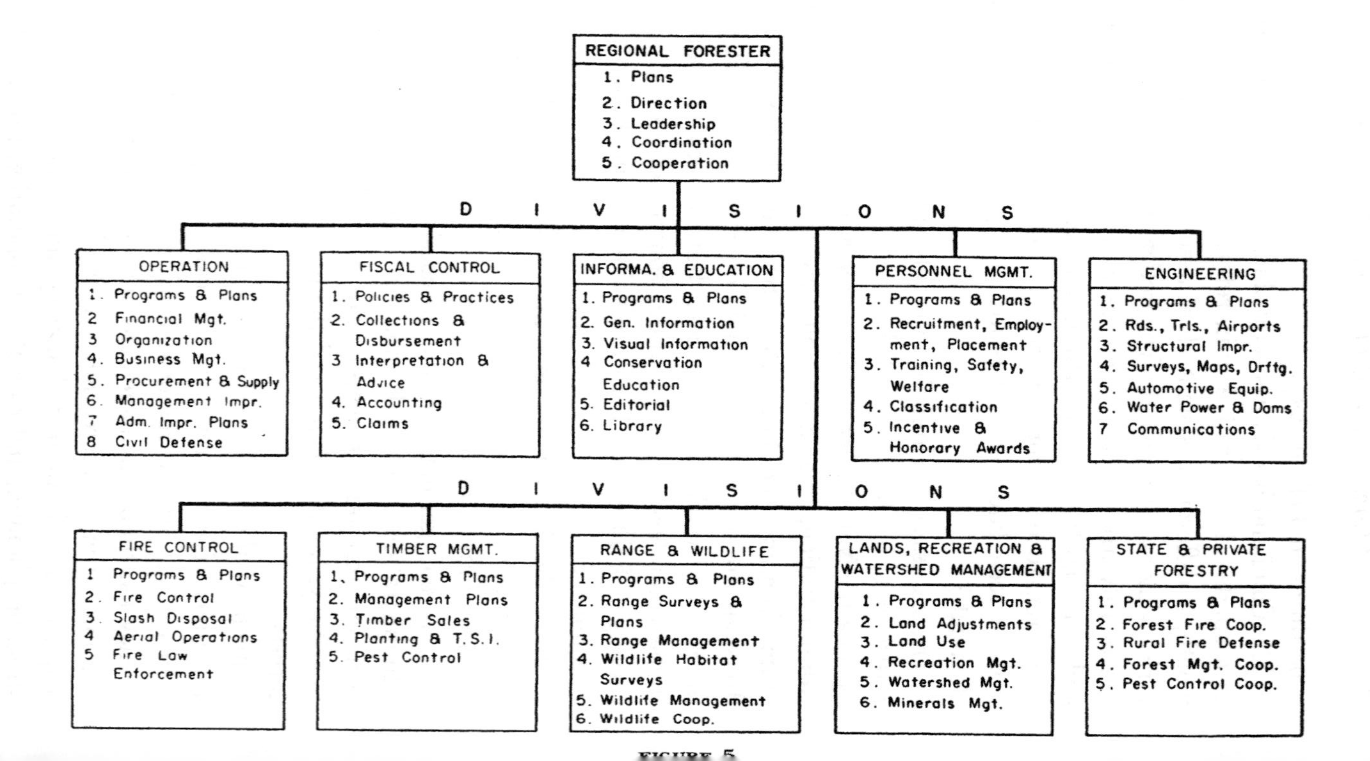
TYPICAL REGIONAL OFFICE ORGANIZATION
REGIONAL FORESTER
1. Plans
2. Direction
3. Leadership
4. Coordination
5. Cooperation
D I V I S I O N S
OPERATION
1. Programs & Plans
2 Financial Mgt.
3 Organization
4. Business Mgt.
5. Procurement & Supply
6. Management Impr.
7 Adm. Impr. Plans
8 Civil Defense
FISCAL CONTROL
1. Policies & Practices
2. Collections & Disbursement
3 Interpretation & Advice
4. Accounting
5. Claims
INFORMA. & EDUCATION
1. Programs & Plans
2. Gen. Information
3. Visual Information
4 Conservation Education
5. Editorial
6. Library
PERSONNEL MGMT.
1. Programs & Plans
2. Recruitment, Employment, Placement
3. Training, Safety, Welfare
4. Classification
5. Incentive & Honorary Awards
ENGINEERING
1. Programs & Plans
2. Rds., Trls., Airports
3. Structural Impr.
4. Surveys, Maps, Drftg.
5. Automotive Equip.
6. Water Power & Dams
7 Communications
D I V I S I O N S
FIRE CONTROL
1 Programs & Plans
2. Fire Control
3. Slash Disposal
4 Aerial Operations
5 Fire Law Enforcement
TIMBER MGMT.
1. Programs & Plans
2. Management Plans
3. Timber Sales
4. Planting & T.S.I.
5. Pest Control
RANGE & WILDLIFE
1. Programs & Plans
2. Range Surveys & Plans
3. Range Management
4. Wildlife Habitat Surveys
5. Wildlife Management
6. Wildlife Coop.
LANDS, RECREATION & WATERSHED MANAGEMENT
1. Programs & Plans
2. Land Adjustments
3. Land Use
4. Recreation Mgt.
5. Watershed Mgt.
6. Minerals Mgt.
STATE & PRIVATE FORESTRY
1. Programs & Plans
2. Forest Fire Coop.
3. Rural Fire Defense
4. Forest Mgt. Coop.
5. Pest Control Coop.

FIGURE 5

management), personnel management, information and education, state and private forestry, and fiscal control. Each of the functions is under an assistant regional forester (except the last, which is under a regional fiscal agent), but several may be combined under one if the workload of the region in those activities is light; no region actually has eleven assistant regional foresters (including one for watershed management), but most have eight or nine. Like the assistant chiefs, they are said to act for their bosses in whatever they do, and each has a staff.

The regions are divided into national forests headed by forest supervisors, whose responsibilities encompass all the foregoing functions but state and private forestry. A few of the larger and busier national forests have assistant supervisors, and a handful even have two. Assistant supervisors are generally qualified to perform as many functions as the supervisors, and assist them by taking up as much of the supervisory load as the supervisors are not able to carry. Whether or not a national forest has an assistant supervisor as well as a supervisor, however, it is sure to have a number of staff assistants to the supervisor who are called "line-staff officers" in the Forest Service; these are held to be different from "pure staff" in that they are explicitly given supervisory authority and responsibility for one or more functions assigned to them, and are not confined in theory or practice to an advisory capacity only. Nevertheless, they are said to act "in the name of" their chiefs. The number of "line-staff" officers on any forest and the number of functions assigned to each depend on the character of the work load; there may be as many as five on some of the more active forests, as few as two on others, and some have assistants.

Every national forest is divided into Ranger districts—some as few as two, others as many as eleven. These are the smallest geographical subdivisions in national forest administration, and the district Rangers who head them are the

TYPICAL NATIONAL FOREST ORGANIZATION CHART

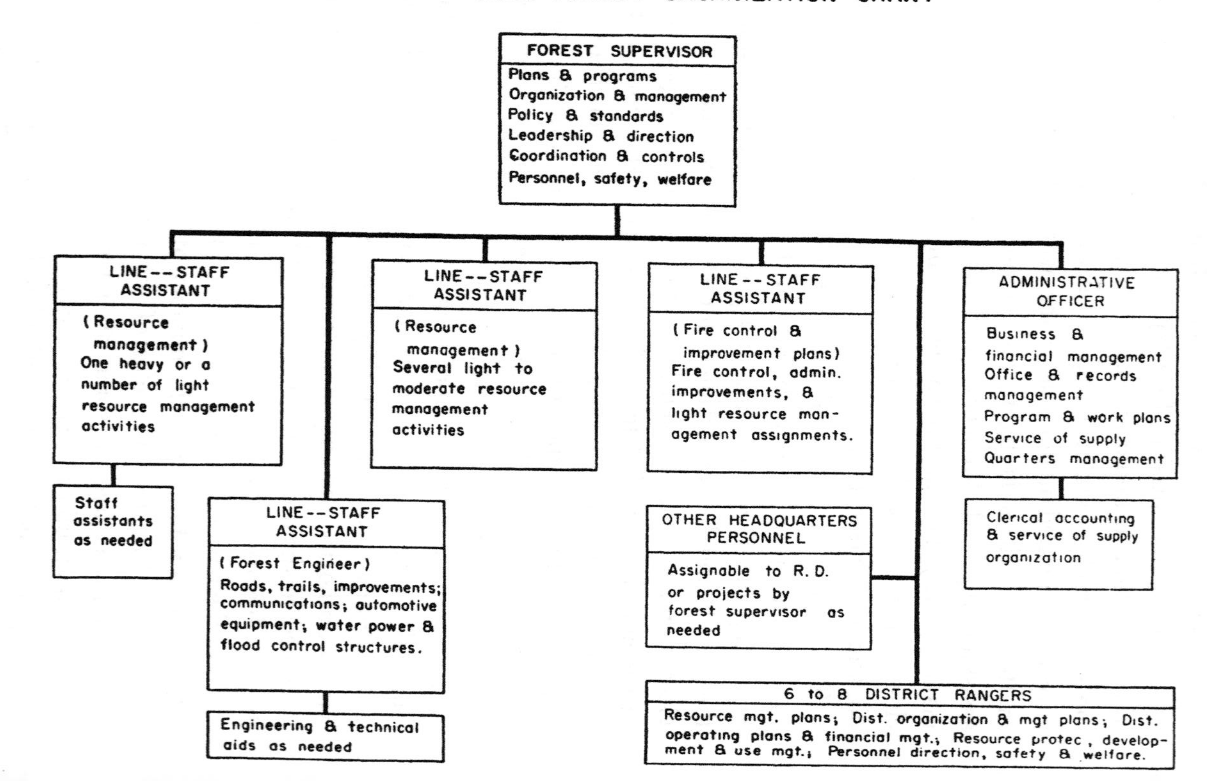
FOREST SUPERVISOR
Plans & programs
Organization & management
Policy & standards
Leadership & direction
Coordination & controls
Personnel, safety, welfare
LINE--STAFF ASSISTANT
(Resource management)
One heavy or a number of light resource management activities
Staff assistants as needed
LINE--STAFF ASSISTANT
(Resource management)
Several light to moderate resource management activities
LINE--STAFF ASSISTANT
(Fire control & improvement plans)
Fire control, admin. improvements, & light resource management assignments.
ADMINISTRATIVE OFFICER
Business & financial management
Office & records management
Program & work plans
Service of supply
Quarters management
Clerical accounting & service of supply organization
LINE--STAFF ASSISTANT
(Forest Engineer)
Roads, trails, improvements; communications; automotive equipment; water power & flood control structures.
Engineering & technical aids as needed
OTHER HEADQUARTERS PERSONNEL
Assignable to R. D. or projects by forest supervisor as needed
6 to 8 DISTRICT RANGERS
Resource mgt. plans; Dist. organization & mgt plans; Dist. operating plans & financial mgt.; Resource protec, development & use mgt.; Personnel direction, safety & welfare.

lowest-ranking professional officers commanding administrative units. The Rangers are responsible for all the basic functions of national forest administration applicable to their respective districts save large construction or other special projects. Most Rangers have one or more assistant Rangers to help them, and organizations of sub-Ranger personnel (including casual, seasonal laborers), but there is no function in which they are not personally active; division of supervisory and planning work is accomplished at this level by dividing up the time of the Rangers, not by employing staff specialists as higher headquarters must.

It takes some 11,000 full-time, permanent employees—almost half of whom are professionals, most of them foresters, many of them specialized technicians—as well as a temporary force of up to 16,000 seasonal and part-time employees to man this organization. Dispersed as they are, holding them on a common course of policy would be a managerial challenge even if they were all doing identical work. Functional and territorial specialization, as well as status differences, tend to heighten the challenge.

The Pivotal Role of the Rangers: Executives, Planners, and Woodsmen

What makes it even more of a trial, however, is that so much of the work by which the objectives of national forest administration are accomplished depends heavily on the Rangers. They actually perform physically the duties of resource management, or directly supervise their physical performance. In the last analysis, the elaborate overhead structure of the Forest Service has as its purpose controlling the behavior of these men who handle the real property of the agency, and who are in most frequent and often close contact with the public.[10] Their tasks require the exercise of judgment; with 792 men making judgments critical to the execution of policy pronouncements, such control is not easily exerted.

FIRE CONTROL

Take, for example, the important function of fire control, which includes prevention (elimination or reduction of fire risks and hazards), presuppression (arrangements made in advance to facilitate suppression when fires occur in spite of preventive measures), detection and patrol, and suppression (actual fire fighting). No other function is as crucial as this, for a forest fire that gets out of hand may wipe out in a single, catastrophic blow all the surface resources spread over hundreds of miles. Insects, disease, weather—these, too, are foes of the forests. But fire alone, if uncontrolled, is so swift and complete. And in the front lines of defense against it stand the men of the Ranger districts.

A substantial portion of the burden of fire prevention is borne by higher headquarters. In particular, mass advertising on a national and regional scale is managed at these higher levels; there is a vast program of publicity designed to educate the public and thus reduce the principal cause of fire: human carelessness. But equal (and perhaps greater) emphasis is placed on locally conducted campaigns of education and persuasion, built around personal contact. The bulk of this responsibility falls on the Rangers.

Thus, Rangers commonly make (or share with their assistants the burden of making) as many as a couple of dozen speeches to local groups each year. Rotary, Kiwanis, Lions, the Grange, fish and game clubs, garden clubs, 4-H clubs,

[10] For a highly informal view of the Ranger's job on a large Rocky Mountain District, see R. McConnell, *Never Marry a Ranger* (New York: Prentice-Hall, 1950).

the Future Farmers of America, colleges, high schools, and even grade schools—in short, every group that might help diminish the incidence of fire—are reminded of the problem; often, slides, film strips, and motion pictures commissioned by the Forest Service are used to enliven the presentations. Community leaders, such as the heads of civic service organizations, school principals, college deans, and managers of large industrial and commercial establishments, are contacted individually whenever this is feasible, and their active support of the program enlisted. Once or twice a year, groups of them are taken, along with local newspapermen and radio station personnel, on "show-me" trips throughout Ranger districts to drive home the imperative nature of precautions against fire (as well as to acquaint them with the work of the Forest Service).

It is also customary for Rangers or some of the personnel under them to visit personally each year a number of the local residents (since there are often many privately owned plots of land within the boundaries of the forests, especially in the East and South and Midwest, and residents dwelling or doing business on government property under permit) and to talk with many summer residents and visitors to public picnic and camp grounds; over several years, most of those who use the national forests regularly are thus contacted individually. (On one district, letters requesting co-operation are sent to a list of 600 residents at the start of each fire season, and individual letters of commendation and gratitude go to persons especially helpful each year.) Warning signs and posters are strategically placed throughout each district.

Rangers also try to maintain cordial, personal relations with local newspaper editors and publishers and heads of local radio stations. That is, in addition to nationwide or regionwide publicity, there is publicity tailored to the requirements of the individual districts. In part, this is obtained by buying space for official notices. Far greater, however, is the amount of donated and editorial space and time made available. One Ranger, for example, has regularly helped newspapers with the drafting of articles and editorials, persuaded local woods-using industries to devote some of their advertising to fire prevention, prepared press releases to win a place in the news columns, and once even arranged to be interviewed by a local radio reporter while on the fire lines battling a small blaze. All Rangers do not lay equal stress on this technique or enjoy equal success with it, but all use it to some degree.

Education, while clearly a mainstay of the fire prevention program, is only the beginning of it. All Rangers, in degrees varying according to their local situations, also allocate substantial proportions of their time to checking fire risks and hazards and eliminating fire trespass. Locomotives and incinerators are inspected for spark-arresting equipment. Logging operations are checked for compliance with fire regulations, including those on the disposal of slash (i.e., of the tops and limbs trimmed off fallen trees). Recreation areas are examined for potential sources of fire. Garbage dumps and other refuse facilities must maintain appropriate safeguards. In some places, during fire season, before people owning property within the gross boundaries of a district may start fires on their own land (which they do to clear the land for planting, to get rid of trash, etc.), they must obtain burning permits from their Ranger; when fires are started in violation of regulations, the Ranger is expected to put an end to these practices, to recover for the government the cost of suppression if the fires get away from those who start them, and even to initiate steps to identify and prosecute willful fire trespassers.[11] Fire prevention calls for the skills

[11] During the depression of the thirties, fires were often deliberately set by people hoping for emergency fire-fighting jobs. Incendiarism is also a way some people express hostility toward the Forest Service. And some neurotic persons start blazes for sheer excitement.

of the policeman as well as of the teacher and the press agent.

It calls for planning, also. Authorized fires must be plotted on accurate maps to avoid unnecessary mobilization of fire-fighting forces. The location of timber sale areas and logging operations must be carefully identified. The frequency, causes, and locations of smokes are recorded and analyzed to aid in the formulation of prevention, presuppression, and suppression strategies. Even locations of warning signs and posters are noted. Fuel types are mapped in detail. Everything that can possibly be employed to develop plans for reduction of the incidence of fires is brought into service by the Rangers.

Yet for all these precautions and tactics, fires do break out —as many as 12,000 covering several hundred thousand acres, in a bad fire year. It is impossible to eliminate them entirely: Cigarettes are flipped casually from car windows despite all efforts to alert people to the risks. Glowing embers of partially extinguished campfires may smolder for hours and finally ignite dry brush and grass. A fire begun to clear a plot of private land may get away from the person who starts it. Lightning may set off a tree. Power saws and other mechanical logging equipment sometimes cause fires. Consequently, no matter how intensive the preventive phase of the fire control program, every Ranger must maintain an organization prepared to detect fires in their early stages (when they are easiest to suppress) and be ready to move promptly into action to meet the threat whenever it occurs.

For detection, the Rangers depend on their lookouts (men posted in towers during fire weather, when low humidity and low precipitation combine with high winds and high temperatures at a time of year in which forest fuels are generally dry), on their wardens or fire crew leaders, on the general populace, and, in special situations, on aerial or ground patrols. From all these sources, information is fed to district headquarters, and men and equipment can be dispatched at a moment's notice to quell any blazes. Seeing that the towers are in good condition, supplied with necessary equipment, furnished with reliable means of communication, and manned by trained personnel at the proper times, are responsibilities of the Rangers. It is also up to the Rangers to see that local people know what to do if they spot the beginnings of a fire too small to be discerned by the regular warning organization. Slip-ups here could be disastrous.

If the detection organization is in good shape, most fires will barely have begun before they are seen and reported. The speed with which the fire organization on any district responds depends on the effectiveness of the presuppression measures taken. Presuppression describes the arrangements that must be made in advance so that men and equipment are available as needed when fires are detected, and so that these can be rushed to the site of the burn to battle it.

Extinguishing a blaze is not an isolated action but the culmination of a long process, a process in which the Rangers have the key roles. Each fire warden or fire crew leader is the head of a team of about ten men who can be summoned to action when the need arises; in the eastern parts of the country, where the warden system is employed, it is up to the Rangers to see that every crew is at full strength, that the wardens can reach them on a moment's notice during fire season, and that every man knows what he has to do. The locations of logging and road crews must be spotted, and the crews prepared for fire fighting, so they can be pressed immediately into service as needed. Emergency crews are organized and trained in local colleges and high schools by Rangers, and Rangers near military installations sometimes arrange to borrow manpower for critical situations. And, since arrangements are made also with other governmental agencies—federal, state, and local—to bring their men and equipment into play at times, and to reciprocate when fires break out on the borders of the national forests and threaten the national forests, all of the regular employees on the Rangers' staffs must be trained both to take their places on the fire lines, if necessary, and to see that fire fighters receive equipment, food, and communications as required.

It may take thousands of men to battle fires that get out of hand; the Rangers are expected to keep the local population sufficiently informed and concerned to volunteer to join the fight when called upon. (Rangers and other district personnel are often sworn in as state forest or conservation officers, and empowered under state law to impress men into service if necessary; while these powers are not often invoked, the Rangers must be prepared to use their authority should the need ever arise.) The Rangers must also be able to identify specialists who can operate relatively complicated pieces of machinery; quality as well as quantity is important.

The men called to action must be able to reach fires quickly, and must be furnished with the tools and other equipment they need to snuff out each blaze. The Rangers are responsible for having the proper implements and apparatus, for keeping them in shape for instant use, and for locating them where they can be brought into play with a minimum of delay. The locations of tools and toolchests, of plows and pumpers and tankers, and of small portable pumps must be picked carefully, marked plainly, and held in readiness and in easy accessibility. Telephone lines and radio receivers and transmitters must be in operating condition, and alternative methods of communication planned in advance should the ordinary ones be put out of commission for any reason; in the Eastern, Southern, and Lake States regions, agreements with commercial telephone companies add their services to those of the Forest Service. Roads and trails must be maintained to permit rapid deployment of men and materials, and slash from logging operations must be disposed of properly by the cutters to facilitate fire fighting. The Ranger who misjudges any of these factors in the presuppression phases of fire control may find small fires raging out of control as a result.

The men on the fire lines must be fed and sheltered. It is up to the Rangers to see that they are. East of the Mississippi, this may mean nothing more than bringing coffee and sandwiches to them; the men designated by the Rangers as fire dispatchers (central communications operators) handle this, placing orders with local storekeepers and restaurants with whom prior arrangements have been made. As crews are relieved, they return to their homes for rest, and reassemble if the fire is still not under control when their next tour of duty comes around. In the West, where distances are too great and fire fighters too scattered to permit such a mode of operations, fire camps are set up. In such cases, merchants and food suppliers are furnished with lists of provisions in advance, and they simply put the packages together and turn them over to truckers for delivery to the camps, where they are cooked and served in camp style. Equipment—tents, blankets, field kitchens, and the rest—must be on hand for hurried establishment of the camps. Everywhere in the country, this requires a good deal of planning beforehand; no matter how populous or accessible the area, there is no time to go shopping when a big fire breaks out.

Prevention and presuppression generally keep most fires from reaching large proportions, but, inevitably, some burn thousands and even tens of thousands of acres. The Rangers are in charge of every fire-fighting operation—big or small—on their respective districts. Officials from their supervisors' offices and regional offices may be rushed in to help, but the Rangers remain as fire bosses regardless of who else is present unless they are officially relieved by a line superior; rarely does this happen—never to any of the Rangers interviewed in this study. The Ranger is responsible for mobilizing and deploying manpower and equipment, and for working out the strategy and tactics of battling the blaze; no one can issue orders to the men under his command or countermand any of his orders as long as he is officially in charge. Fire is an unpredictable and a ruthless enemy; with a strong wind behind it and a highly flammable fuel type (such as resinous pine), it may overtake a speeding truck, and it can change direction without warning. Flames may reach heights of 60 to 100 feet, leap over a line of fire fighters and start new fires behind them.

The Rangers are responsible for the safety of the men under them, men who may be scattered over miles of territory, and they must often make split-second decisions, weighing the threat to the forests, to private property in the path of the fire, and to the fire fighters. It is hot, heavy, dirty, nerve-wracking labor, and the way the Rangers manage it will generally make the difference between a disaster to life and property and a moderately damaging burn. As long as a big fire goes on, a Ranger is likely to go with little sleep or rest; it is always his responsibility.[12]

When the fire is under control, the Ranger's work is not over. Emergency crews are released, but the regular Forest Service employees, and some of the wardens and their standby crews, continue mopping up until the last ember is extinguished. Then, at last, the weary task of reassembling equipment and preparing it for the next emergency, and, eventually, the filing of reports. There are investigations to determine the causes, to assess the damages, to evaluate the strategy, to plan the regeneration of the area burned over. Fire fighters, truckers, food suppliers, equipment providers submit their bills for their services and material, all of which must be examined, adjusted, approved, and submitted to higher headquarters for payment. Obviously, cost of extinguishment is a secondary consideration, but it is up to the

[12] For a fictional, but highly accurate, account of the details of fighting a large fire in the West, see G. R. Stewart, *Fire* (New York: Random House, 1948).

Rangers to exercise discretion to avoid gross extravagances and payment for services not actually rendered.

No one will deny the importance of the planners of the grand, nationwide strategies, the formulators of the great designs and broad policies. But they know, as every Ranger knows, these all rest ultimately on the performance of the men in the field. In the conquest of fire, in the management of the resources of the national forests, practically everything depends on the Rangers. In their behavior lies the secret of success or failure.

TIMBER MANAGEMENT

As in fire control, success or failure in timber management lies in the behavior of the Ranger. Wood is a valuable commodity, and literally hundreds of thousands of dollars worth stands in every Ranger district. Paper manufacturers seek it for pulping; lumber processors want it for boards and ties; furniture makers need it for veneer; farmers ask for it to build houses and barns and fences; it is sought for firewood and for Christmas trees. All five Rangers try to sell the allowable cut for their respective districts, while at the same time satisfying as many of the potential buyers as possible and getting the best prices for the government. Later on, it will be seen that the forest supervisors and the regional offices are drawn into the negotiations for larger sales, sometimes only formally, often quite actively. On the whole, however, the Rangers and the personnel under them, knowing more about the status of the timber under their control than anyone else, and in closer touch with the demands for the timber they can offer for sale than anyone else, play the chief roles in the management of the timber resources under their respective jurisdictions.

Thus, for example, within general sales policies of the Service and within the limits of silvicultural requirements, the judgment of the Rangers counts heavily in the determination of the size of a safe cut in terms of sustained yield, of the areas to be cut, of the terms of the agreements with the purchasers. Rangers skillful at bargaining may persuade an operator to take out some less desirable species as well as more valuable types of tree in order to prevent encouragement of the less valuable types in the areas cut over; others, anxious not to jeopardize an attractive bid, may prefer to let the loggers take only the economically preferred trees and to send Forest Service employees into the sale area afterwards to girdle the unmerchantable species. Some are inclined to promote large numbers of small sales to local people; others foster large sales to large companies, whether local or not, in order to take advantage of the lower overhead costs incurred by such transactions, to facilitate planned cutting, and to simplify the problems of policing the terms of the contracts. Some are aggressive in pushing sales; others more inclined to let economic demand registered through the marketplace guide their programs.

Further, some Rangers try to discourage bidders from driving prices so high above the levels considered acceptable to the government as to make profitable processing impossible (a practice that loggers desperate for wood will not infrequently engage in, sometimes hoping for sharp rises in the prices of finished products, sometimes seeking nothing more than a winter's work for labor income alone, there being few alternative employment opportunities in many forest communities); others let the bidders run their own risks without advice. Some prefer to require loggers to build timber access roads (to Forest Service standards), making due allowance for this additional expense in stumpage price determinations; others would rather have roads built by the Forest Service and paid for by the timber operators' deposits with the agency of amounts equal to the costs. Some construe sales contracts narrowly and rigidly in every detail (which means added costs to the buyers); others tend to be a little more lenient. Some favor lighter thinning and natural regeneration where others favor heavy cutting and extensive planting. In no sense do the Rangers have completely free hands in these decisions (as later discussion will show), but higher levels commonly defer to the Rangers on many matters because their knowledge of their districts is more intimate than that of anyone else.

For Rangers spend roughly half their time in the field. In connection with timber management, field time is devoted to supervising, inspecting, and participating in the silvicultural work. They cruise timber (i.e., inspect trees and estimate the volume of their usable contents) or check the activities of their trained subordinates. They select or inspect the selection by their aides of trees to be designated for cutting. In some parts of the country, it is still common for forest officers and their employees to scale (i.e., measure) logs and pulpwood "sticks" in the woods.[13] Sale sites are planned on the basis of careful field study to determine the best methods of operation, of slash disposal and of stand improvement following the completion of a logging operation. The areas must then be inspected to ensure conformity with the terms of contracts (including the cutting of only designated trees, removal of less desirable species or trees when this is made part of an agreement, proper disposal of slash, and elimination of fire hazards created by logging operations). If planting is required, it takes field surveys to draw up the plan of action and field checks on plan adherence.

Also, free-use permits (for the removal of dead and down

[13] The practice is declining because scaling is expensive, and the difference between estimated and measured volume frequently turns out not to warrant the expense. Increasing reliance is being placed on "consumer scale" (measurement by the consumer under tight controls) and on estimates (with buyers accepting the Forest Service figures, and now managing to saw out a little more than they pay for, now a little less). But differences between estimated and actual volume constitute a challenging and troublesome technical problem; so scaling of logs before removal from the woods persists in many places.

timber in limited amounts for personal use) and personal-use sales (limited to local residents or settlers, who may buy live timber, but may not resell it and must use it for personal purposes, such as the construction of a house or barn) necessitate field work despite their relatively small size. Unauthorized removal of timber, whether deliberate or inadvertent, can be discovered only by constant vigilance in the woods, and it is up to the Rangers to recover for the government the fair market value of the property taken, to try to collect penalties in addition when the trespass appears intentional, and even to carry cases to higher headquarters or to local district attorneys and law enforcement officers for help in the identification and prosecution of offenders on occasion. All these duties compel the Rangers to spend a great deal of their time in the field (which most of them prefer to office work, anyway), and to become as thoroughly familiar with the resources in their charge as they are with their own back yards. They delegate—they are expected, and even *required*, to delegate—to subordinates as many of the routine tasks as they can. But they still have to be in the woods much of the time, and their knowledge of their areas is consequently great even when they are still relatively new to their districts. Hence, not only must higher headquarters rely on the Rangers to get the work of timber management done properly, but the Rangers' assessments of what is "proper" must command respect. The role of the Rangers is thus doubly important.

No two Rangers would be likely to handle the timber resources of a given district in precisely the same fashion. They stress different aspects of the job, and go about it in their individual ways. But, since the Rangers are the experts with respect to their own jurisdictions, it is not a simple matter for those above them to contradict them. Whatever timber policy might mean to those who formulate it at the national level, on the ground it means what the Rangers and their staffs do.

RECREATION AND LAND USES

Similarly, the other phases of national forest administration are contingent upon the administrative behavior of the Rangers. For instance, millions of people want to picnic, camp, hike, hunt, fish, and ski in the national forests. Companies want to extract minerals from them, dam their streams (for power or irrigation), run power lines and pipelines and railroads across them. Local, state, and federal agencies want to build roads through them, put garbage dumps in some parts and schools in others and construct reservoirs in still others. Concessionaires seek to erect lodges, eating establishments, resorts, and sports areas. Civic groups, Boy Scouts, 4-H Clubs, and others, want to set up summer camps. City-dwellers come looking for places to build summer homes. They all generally come first to the Rangers, or, if they go higher in their first approach, are referred to the Rangers.

The Rangers' dispositions of these applications, whether favorable or negative, are not final. But they weigh heavily in the final decisions. For within the standards set by national policy, and even after a comparatively short time on their districts, the Rangers are in a better position than anyone else in the Forest Service to calculate what the lands they manage can safely sustain, what projects are likely to arouse vehement opposition from other forest users, and what enterprises are likely to conflict with possible future developments of their areas. They are also in the best position to suggest the kinds of restrictions and conditions that ought to be included in permits and agreements. And the burden of holding permittees to the terms of permits and agreements falls chiefly on them. Some routinely transmit applications to higher headquarters; some are highly selective, trying to discourage those that would complicate other phases of resource management or create what they consider eyesores;

some actively try to promote development of all the varied uses of their districts; some, with an eye to the future, fight to protect sites for eventual public use instead of supporting requests for personal or restricted or commercial uses.

Furthermore, the process of land acquisition and exchange goes on even though the major outlines of the national forests have been completed.[14] In the East and the South and the central part of the country, where most of the national forests were acquired by purchases and there are still many enclaves of private property that present problems of management and difficulties in relationships with the public, the Forest Service is still endeavoring to consolidate the government tracts, largely through land-for-land exchanges. Even in the West, where it was possible to reserve great blocks of land still in the public domain, the straightening of boundaries and solidification of government holdings by exchange goes on. More often than not, the initiative for a change will come from Rangers desiring to improve the administration and appearance of their districts. In addition, the first approaches by individuals who want to effect exchanges for reasons of their own are made to Rangers. In any event, the opinions and recommendations of Rangers affected by proposed adjustments will invariably be canvassed. One will be aggressive in pursuit of exchanges; indeed, one Ranger interviewed in the course of this study persuaded an applicant for a Forest Service parcel, who had no land to offer in exchange,

[14] Lands may be acquired for national forest purposes (1) by reservation from the public domain (a process now completed, by and large), (2) by purchase, (3) by donation, or (4) by exchange of timber, or of other national forest land within the same state, when this is to the public interest.

East of the Mississippi, changes in national forest boundaries must be approved by the National Forest Reservation Commission in Washington.

National forest land may not be sold, except for small tracts whose chief value is agricultural. If such tracts are not needed for public purposes, they may be sold to actual settlers for homesteading.

to buy a parcel the Ranger wanted and to offer that in exchange for the land he sought (and that the Forest Service did not need). Another will not actively promote exchanges, but will process applications if they come in. Still another will discourage them. The procedures and objectives are spelled out in statutes and administrative regulations, but the practices depend on the Rangers.

RANGE MANAGEMENT

In the same way, the Rangers stand out in the management of grazing lands. In the western regions especially, they have struggled to hold down the number of livestock on government property on the basis of their estimate of the carrying capacity of the land; while the decisions on the optimum size of herds are promulgated by higher authorities, it is the Rangers' knowledge of the terrain that forms the basis for these decisions. They play important roles in determining the number of head approved in each grazing permit. They lay out the boundaries of the "allotments" (the areas in which permittees may run their livestock). They assess the readiness of the range to receive stock, and may delay opening it for a time. They make findings as to the "commensurability" of ranchers (the private holdings of permittees must be sufficient to sustain their herds when not on national forest range, ensuring that grazing privileges go to legitimate ranchers instead of speculators), and they can effectively prevent sales of herds or ranches (with which go preferences on the national forests) by reporting to the prospective sellers and purchasers that the latter are not "commensurate" (or are for some other reason unacceptable) and therefore ineligible for permits. They must police permits, making sure that the livestock actually run on the ranges are not in excess of the number allowed, and they must see to it that the animals do not stray from their allotments to the allotments of neighboring permittees. They are responsible for promoting revegetation of the ranges, and for fencing them where necessary. They must balance the demands of grazers with the demands of other forest users—with recreationists, who call for camping grounds and other facilities that often conflict with grazing; with hunters, who favor the encouragement of game herds as against cattle and sheep; and with others. General policies are announced from Washington, but the Rangers are so obviously influential in adapting the words to local conditions and applying the adapted policies to the range that the chief of the Forest Service has publicly acknowledged it.[15]

OTHER FUNCTIONS

And so it goes with the other functions in resource management. Rangers work with state conservation agencies in wildlife management and study wildlife habits and conditions at first hand; their recommendations as to hunting seasons and the accommodation of hunters, the establishment of game refuges and forage areas, bag limits, etc., carry great weight in determinations of the content of wildlife programs. Water management is, on the whole, a by-product of good forest management, but many Rangers take special pains to safeguard reservoirs in their districts from damage by loggers and grazers, and to protect the headwaters of rivers in the high country. Information and education programs at the district level concentrate on fire prevention, as noted earlier, but not to the exclusion of opportunities for presenting explanations and justifications of Forest Service policies in other functions; like many another organization, the Forest Service has found community public relations an essential supplement to massive national campaigns.

Personnel management in the districts is primarily a task

[15] See pp. 190-191.

of recruiting "wage board" employees (i.e., employees whose wages are set by departmental boards rather than by the Classification Act) and sub-professional aids, gathering the data on which wage boards base their rates for laborers, and, most important of all, training such employees to do manual work in the field. (Some districts have as many as eighty employees during the summer.) Professional personnel are recruited by the Washington office through the Civil Service Commission, but their major training is administered by the Rangers to whom they are assigned; in fact, the training they get in the early years before they are moved to districts of their own is unquestionably the most intensive, and perhaps the most important, to which they are ever exposed in the Service. Finally, under the rubric of General Administration, the Rangers prepare reports, keep up with their official correspondence and records, read the heavy volume of directives and circulars from higher headquarters, and maintain (or supervise clerical maintenance of) the files and manuals that give continuity to administration in their units.

In short, when people talk of the accomplishments of the Forest Service, they are speaking in large part of the way the Ranger districts are managed by these executives in the field. As resource managers, with heavy responsibilities and considerable opportunity to exercise their own discretion, they could conceivably go off in many directions, running their districts in widely varied and totally unrelated fashions. The more the Rangers are given to do, the greater the risk of such dissolution. Yet the more intensive the administration of the national forests, the more responsibilities, inescapably, must be assigned to the Rangers.

The Thrust toward Disunity

The Forest Service was born late, grew fast, has a broad range of functions, and a large and finely divided organization. Leading an agency under these conditions, especially in the face of strong and carefully marshalled opposition from politically powerful interests, is not an easy task; the leaders have many things to watch, and respond to, simultaneously.

In addition, they must accept their heavy dependence on their field officers. The substantive content of the agency program is shaped by what the men in the woods do from day to day. Not only does the conduct of each function reflect the field men's interpretations of their jobs, but the balance among functions grows out of their activities and decisions. By emphasizing one function over others, by aggressiveness or passivity, by inventiveness or adherence to the *status quo*, by risking the displeasure of superiors or colleagues or neighbors or by following the path of least resistance, by enthusiastic or indifferent or reluctant performance, the Rangers in effect modify and even make policy—sometimes without knowing it.

These factors provide strong thrusts toward disunity. Yet these are not the only such forces; the next chapter takes up many more of a different, though related, character.

CHAPTER III

CHALLENGES TO UNITY

Problems of Internal Communication

GENERAL INSTRUCTIONS VERSUS SPECIFIC SITUATIONS

Necessarily, the federal statutes under which the Forest Service operates are framed in general terms. This applies to the more than eighty laws applicable to the Service and the national forests specifically as well as those that govern the whole federal establishment—laws on personnel, budgeting, purchasing, contracting, and other staff services.

Similarly, the directives of the Secretary of Agriculture and the instructions from the Washington office of the Forest Service are couched in rather sweeping terms, for they must be drafted in sufficiently broad language to apply to all the different conditions obtaining in the far-flung national forests. Even the issuances of the regional and supervisors' offices cannot be too specific, for the regulations of the smallest region apply to seven national forests, and the largest region encompasses twenty-six, while most national forests contain four to six Ranger districts, and one has eleven.

Consequently, unless higher authorities were disposed to write individual instructions for each Ranger district—which would require them to know the details of the situation on each field unit, a manifest impossibility—they would seem to be compelled to allow considerable latitude for independent judgment on the part of the field officers with regard to the interpretation and application of the more general statements. Under these circumstances, personal and local factors assume substantial proportions in the management of the districts, increasing the likelihood of significant variations in policy as it emerges concretely on each such unit.[1]

INCONSISTENT DIRECTIVES

Furthermore, the instructions field officers are called upon to execute are not always easily reconciled. While the Forest Service, as seen by the Chief or by the casual outside observer, assumes the form of a traditional administrative pyramid, it appears to the individual Ranger as an inverse pyramid with himself at the apex. For as he looks upward at the organization from his position in the field, he sees a forest supervisor aided by several staff assistants and perhaps by an assistant supervisor, all of whom send him communications and instructions (frequently over the name of the supervisor, but often, too, over the assistants' signatures). Above them is a regional forester with a more numerous and narrowly specialized group of assistant regional foresters with aides of their own in their divisions; the Rangers know that a good deal of what reaches them through the supervisors and staff assistants originates with these specialists. At the top, the organization burgeons out into the elaborate, highly

[1] For an interesting illustration of this problem, see J. Hersey, *A Bell for Adano* (New York: Alfred A. Knopf, 1944), pp. 13-15. Here a civil affairs officer on duty in a small town in Italy during World War II "took the sheets of instructions [from higher headquarters] up from his desk and tore them in half, tore the halves into quarters, and crumpled up the quarters and threw them into a cane wastebasket under the desk. . . . Plans for this first day were in the wastebasket. They were absurd. Enough was set forth in those plans to keep a regiment busy for a week."

specialized branches and divisions serving as staff to the Chief. To switch metaphors for a moment, the Forest Service seen from the level of the district Ranger looks like a vast funnel with the Ranger at the throat of it; all the varied elements and specialties above him pour out materials which, mixed and blended by the Ranger, emerge in a stream of action in the field.

What creates a problem for the field man is the fact that the materials sometimes require mutually exclusive courses of action on his part. The specifications for roads, for example, are generally predicated on engineering premises alone, but roads built to those specifications may conflict with the demands of watershed management or recreation management or timber management specialists. Rigid adherence to timber management or range management program goals can generate public criticism that provokes objections from information and education specialists. Emphasis on recreation that gratifies recreation officers may disturb fire control officers. What looks like adequate concern for the grazing uses of the national forests may seem like indifference to wildlife management from the perspective of those who specialize in this function. Administrative assistants call for greater attention to office routines and procedures and paperwork, while other functional specialists deplore expenditures of Ranger time in the office rather than in the woods. Despite the general consensus on the desirability of multiple-use resource management, it is not always clear what this indicates in specific instances.

The problem appears to be endemic in large-scale organizations.[2] It means field men must resolve conflicts in their

[2] See, for example, how it appears in a work relief agency in A. W. Macmahon, J. D. Millett, and G. Ogden, *The Administration of Federal Work Relief* (Chicago: Public Administration Service, 1941), Chapter 11; in the conduct of foreign affairs in A. W. Macmahon, "Function and Area in the Administration of International Affairs," in L. D. White, et al., *New Horizons in Public Administration* (University of Alabama

own ways. As a result, there is always a strong possibility that there will be not one, but many policies.

DISTANCE

Distance is used here in two senses: sociologically, to describe status, linguistic, and attitudinal barriers that divide components of organizations; and physically, to refer to the geographical spaces that separate members of organizations from one another.

Status distance—essentially, differences in rank and in the deference accorded people of different rank—operates as a devisive element in many organizations by filtering and distorting communications upward and downward.[3] On the one hand, people sometimes tend to tell their superiors only what they think their superiors want to hear: "most executives are very effectively insulated from the operating levels of the organization." Moreover, many leaders are not receptive to ideas put forth by their subordinates: "a suggestion coming from a person of low status will not usually be treated with the same respect and seriousness as one coming from a person of high status." On the other hand, casual remarks, inquiries, and tentative observations of high-status individuals are often emphasized and reinterpreted and applied in ways those who make the statements never actually intended. That is why one expert student of organization talks about the pathology of status systems, and of their disruptive tendencies.[4]

Linguistic barriers often arise as consequences of the unique universes of discourse that spring up in connection with occupational specialties and functional differentiations

Press, 1945); and in the administration of a city health department in H. Kaufman, *The New York City Health Centers* (New York: The Inter-University Case Program, 1952).

[3] H. A. Simon, D. W. Smithburg, and V. A. Thompson, *Public Administration* (New York: Alfred A. Knopf, 1950), p. 236. The quotations in the above paragraph are from the same source.

[4] C. I. Barnard, *Organization and Management* (Cambridge: Harvard University Press, 1948), Chapter IX, especially pp. 231 ff.

in organizations. Special meanings are attached to common words, special vocabularies emerge, special modes of expression evolve. Often, these serve as a shorthand method of communication among the members of each in-group. Sometimes they permit greater accuracy of expression, eliminating ambiguities for those who understand the jargon. Nevertheless, however useful these practices may be for the initiates, they frequently make it difficult for the various specialists to discourse with one another. The specialists use the same words in different ways and completely misunderstand each other, or they are unable to decipher each other's communications.[5]

Attitudinal barriers appear when people have totally different frames of reference. The same facts then appear in different perspectives, and lead to different conclusions. "The stimuli that fall on a person's eyes and ears are screened, filtered, and modified by the nervous system before they even reach consciousness—and memory makes further selections of the things it will retain and the things it will forget." [6] Many things go into shaping mental sets, and no two individuals are likely to be identical in this respect. Moreover, even people who come to an agency with very similar attitudes and values will often develop divergent points of view according to the places they come to occupy and the functional specialties with which they become associated.[7] Thus, there are often wide discrepancies between what members of organizations *intend* to communicate to each other and what they *actually* convey.

Social distance in this fashion can easily lead to the pursuit

[5] Simon, Smithburg, and Thompson, *op. cit.*, pp. 229-32.

[6] *Ibid.*, pp. 232-35.

[7] H. A. Simon, *Administrative Behavior* (New York: The Macmillan Co., 1947), p. 214 n., records a perfect illustration, in which a bureau chief, serving as acting department head for his absent superior, disapproved a number of proposals he himself had initiated as bureau chief. "From up here," he is reported to have explained, "things don't look the same as they do from down there."

of many different policies in an organization allegedly having only one. It produces grave distortions in the communications network, permitting officials throughout the hierarchy to strike out on a profusion of unrelated paths.

If this problem is vexing because of distortions of communication, it must obviously be even more troublesome where spatial distance reduces the frequency and length of contacts. In an agency as scattered as the Forest Service, this is far more than a hypothetical difficulty.

Modern techniques of communication and transportation have shrunk geographical distances, it is true. Few field installations of the Forest Service in continental United States cannot be reached from Washington in twenty-four hours or so, and in proportionally less time from regional and supervisors' offices. Federal, state, and local highways, as well as roads and highways constructed by the Forest Service itself, now make almost every Ranger station accessible by automobile, and virtually all field stations (as well as higher headquarters) are equipped with one or more vehicles for official use. Where commercial telephone lines have not yet reached, the Forest Service has strung its own lines to connect Ranger stations with lookout towers and made other improvements. The Service has been making increasing use of FM radio links (including transmitters and receivers in its vehicles). Planes have been added to the fire detection and suppression facilities, and smokejumpers now arrive in hours at blazes that once would have taken days and even weeks to reach.

Nevertheless, many field stations are inconvenient to reach even though this can be done fairly quickly. If a trip takes the better part of a day, one way, members of the Service will not shuttle back and forth casually; they have neither time nor money for so much travel, and accumulate whatever business must be transacted in person to be handled in as small a number of trips as can be managed. In fact, even forest supervisors do not make a habit of running out to their Ranger districts with every little matter that comes up, nor do the Rangers rush in to the supervisors with every question, for even on a small national forest, a round-trip journey of this kind is apt to consume the better part of half a day. Commercial telephone rates are just high enough to discourage field officers from making toll calls (and thus using up their communication funds) for minor issues, and Forest Service telephone lines generally run from Ranger stations to the field rather than between Ranger stations or from Ranger stations to higher headquarters. (While many national forests employ their radio transmitters to establish communication links among these offices, the range and reliability of FM connections are still limited, and do not seem yet to have spanned the gaps entirely.)

To be sure, the Rangers are far from isolated today compared to what they were only a generation ago;[8] the volume of contacts with superiors and with other Rangers has unquestionably increased many times in recent years. Yet they are still forced to make many decisions and resolve many doubts without reference to those above them; they are thrown back on their own resources to a larger extent than most bureaucrats. As a means of co-ordination, "the device of propinquity, the juxtaposition of offices in the same building or city, and reliance on ordinary daily contact," [9] is for all practical purposes denied the Forest Service. Geographical dispersion is thus a powerful centrifugal factor here.[10]

[8] Cf. pp. 84-85.

[9] L. Gulick, "Notes on the Theory of Organization," in L. Gulick and L. Urwick (eds.), *Papers on the Science of Administration* (New York: Institute of Public Administration, 1937), p. 36.

[10] That distance is still a factor in communication in the Forest Service is indicated by the general agreement among the personnel interviewed that the more accessible forests and districts are more frequently visited and inspected by representatives of higher levels than are the more remote areas. For this reason, some field men prefer the more distant stations.

Behavioral Norms of Face-to-Face Work Groups

A voluminous literature has grown up in recent years describing the characteristics of small groups in face-to-face contact—"primary" groups.[11] Among the results of these studies is the accumulation of strong evidence that groups of this kind tend to develop norms of their own, norms governing procedures and relationships among personnel and work output, to which the members adhere. In general, the norms evolve without conscious or systematic effort on anyone's part; indeed, it sometimes appears the individuals controlled by them could not articulate them if they were asked to, although the observer can see quite plainly that the standards determine what personnel do and decide. Unconsciously, the members act in the group-prescribed way because that is simply "the way things are done." Sometimes, it is true, those who violate these informal standards are subjected to sanctions of many kinds by their fellows. Most of the time, however, the norms take effect without many people being at all aware that the process is operative.

If the leaders of an organization are fortunate, these informal norms will coincide with formally promulgated regulations. Often, however, conflicts develop. Output is then restricted by the workers, methods are adopted that are at odds with prescribed procedures, changes in modes of operation or in established relationships are resisted.[12] What work

[11] Comprehensive surveys of this literature are contained in A. P. Hare, E. F. Borgatta, and R. F. Bales, *Small Groups* (New York: Alfred A. Knopf, 1955); *The American Sociological Review*, Vol. XIX, No. 6 (December, 1954); D. Cartwright and A. Zander, *Group Dynamics* (Evanston: Row, Peterson and Co., 1953); and G. C. Homans, *The Human Group* (New York: Harcourt, Brace and Co., 1950). Of particular relevance to the discussion here is Part Three of the Cartwright and Zander book.

[12] F. J. Roethlisberger and W. J. Dickson, *Management and the Worker* (Cambridge: Harvard University Press, 1939), Chapters XXII, XXIII, XXIV; S. B. Mathewson, *Restriction of Output among Unorgan-*

groups actually do may consequently be quite different from what the leadership intended or anticipated.

This presents a challenge in the Forest Service because most Ranger districts employ non-professional workers, under the Rangers and assistant Rangers, who remain on the districts for many years, in contrast to the professional men, who change frequently because of Forest Service transfer and promotion policies (discussed in later chapters). These permanent personnel are usually not many in number, varying from one or two on districts having short field seasons to as many as ten or fifteen on districts where long field seasons justify the maintenance of a year-round work force. (Temporary employees engaged as seasonal labor fill out the work force during the months of mild weather.) But they come to know the districts and the local people intimately, and the frequently-transferred executives over them depend heavily on them in the management of district resources and programs.

Considering what sociologists have found in their studies of large-scale organizations and of small groups, it might be expected that the local standards would gradually impress themselves on the Rangers, that the norms of conduct prevailing locally would occasionally clash with the official requirements of the agency, and that some Rangers would consequently deflect in some respects from those requirements. And there is some evidence that they do—for example, in whether or not they wear the Forest Service uniform

ized Workers (New York: The Viking Press, 1931); B. B. Gardner and D. G. Moore, *Human Relations in Industry* (Chicago: Richard D. Irwin, Inc., 1950), Chapter 18 and pp. 372 ff; B. M. Selekman, "Resistance to Shop Changes," *Harvard Business Review*, Vol. 24, No. 1 (Autumn, 1945), pp. 119-32; R. N. McMurry, "The Problem of Resistance to Change in Industry," *Journal of Applied Psychology*, Vol. 31, No. 6 (December, 1947), pp. 589-93; L. Coch and J. R. P. French, Jr., "Overcoming Resistance to Change," in S. Hoslett (ed.), *Human Factors in Management* (New York: Harper and Brothers, 1951), pp. 242-68 (especially at pp. 265-66).

regularly, insist on formal training sessions for their subordinates, or follow work-planning "guidelines" to the letter, the Rangers seem to be influenced by habits and reactions of local groups of employees. (National forests as entities also tend to fashion rather special ways of doing things; one, for example, handled its budgeting in a way quite at variance with regulations, a departure from prescribed practice no one on the supervisor's staff or among the Rangers realized had occurred or thought to check; the "common-sense" procedures evolved on that forest simply became the customary and accepted method of financial planning.)

The formation of "informal" groups with parochial patterns of behavior sometimes in conflict with official regulations is a phenomenon with which relatively compact organizations are compelled to deal. The problem in an agency as scattered geographically as the Forest Service, whose component face-to-face groups are comparatively isolated, is especially acute, and constitutes a particularly serious challenge to the cohesion of the organization.

"Capture" of Field Officers by Local Populations

"Where considerable authority is devolved upon field officials," observes one eminent student of government, "there is always the danger . . . that policy will be unduly influenced by those private individuals and groups who are in closer and more intimate contact with the field than are the superior officers . . . Localized influence, . . . if carried to any great lengths, is likely to beget such differences of policy between field offices that national policy will be a fiction." [13] Clearly, considerable authority is indeed devolved upon the field officials of the Forest Service, and the danger noted is therefore substantial.

[13] D. B. Truman, *Administrative Decentralization* (Chicago: University of Chicago Press, 1940), pp. 14-15.

The danger is really twofold. On the one hand, there is the risk that field men, regarded by their chiefs as emissaries sent to live among local populaces and represent the agency to the people, become so identified with the communities in which they reside that they become community delegates to headquarters rather than the reverse.[14] On the other hand, there is the possibility that the field men, though devoted to their leaders, might be cowed by local pressures.

The first aspect of this danger arises in the Forest Service because so much emphasis is placed on the local roots of the field units;[15] Rangers are encouraged to take as active a part as they can in community service, social, and fraternal organizations. Slowly, they absorb the point of view of their friends and neighbors. One, for example, reported that he found himself tending to "look the other way" and to delay investigation as long as he reasonably could when he had reason to believe the chamber of commerce of the town in which he lived, and to whose executive committee he belonged, was operating a resort area without the rather expensive liability insurance required by the terms of its special-use permit. Another argued vigorously in favor of disposing of timber through many small sales instead of fewer large ones (preferred by his superiors because the large ones are more economical, simpler to administer and police, and make the achievement of management goals easier to attain); though he did not deny the administrative advantages of

[14] The process is described amusingly in J. Patrick, *The Teahouse of the August Moon* in L. Kronenberger, *The Best Plays of 1953-54* (New York: Dodd, Mead and Company, 1954); here, the inhabitants of an Okinawan village succeed in making Okinawan villagers of American officers sent to bring them American culture. In I. Stone, *Lust for Life* (New York: The Modern Library, 1939), pp. 26-83, the same theme is treated tragically as a missionary to a Belgian mining town becomes indistinguishable in outlook and behavior from the miners whose lives he was to enrich through religious teaching and practice.

[15] See the remarks of the Chief of the Forest Service quoted on pp. 190-191.

larger sales, he contended they attract large "outside" companies able to outbid the local operators whose interests he sought to defend.

Also, in a couple of instances, Rangers decided against termination of grazing permits for small herds better removed (from the technical standpoint) from their districts, and of no importance to the local economy, just because they felt such action would work some hardship on the families who owned the cattle. The people the Rangers sought to protect were not influential enough to cause the Rangers any difficulty had the Rangers not fought for their welfare; it was not anxiety of this kind that moved the forest officers. Rather, it seems to have been a genuine concern for the individuals involved, a sense of community with them. Men who thought only in terms of their organizations rather than in terms of their friends and neighbors doubtless would have been less sensitive to the latter's personal needs and desires.

The second aspect of the danger is that many local interests are in positions to bring considerable pressure to bear on the field man. Sometimes it is exerted on him directly—to convince him of the wisdom and justice of the local demand, if possible; to compel him to accede to the demand, whatever his convictions, if necessary. They can visit him repeatedly in his office and home, argue with him at meetings of civic and fraternal associations, talk with him at gatherings of church groups. News reports and editorials in the local press and on the air ventilate local grievances and aspirations. (Incendiarism has been employed in some areas as a coercive measure against the Forest Service, and one Ranger was threatened with violence by an armed man in the backwoods of West Virginia when the Ranger was investigating a possible timber trespass, but these, of course, are not the ordinary modes of persuasion and protest.) Even the most devoted forest officer might understandably yield to these constraints now and then; on a large scale, the results could gravely undermine the unity of agency policy.

Sometimes the pressure is directed over the head of the Ranger and exerted on his superiors. Statutes and regulations of the Department of Agriculture establish regular procedures for the handling of appeals from administrative decisions, and a person affected by a Ranger's action (or inaction) may file formal requests for hearings and redress with the forest supervisor, and carry his case to the regional forester, the Chief, the Secretary of Agriculture, and ultimately to the courts if he is so minded. More commonly, aggrieved individuals simply write informally to higher headquarters and pray for relief; in most cases, this is sufficient to precipitate an inquiry into a field officer's decision.[16] Many people get in touch with their United States Representatives or Senators, whose queries to the Chief on behalf of their constituents ordinarily get prompt attention and cause reverberations throughout the hierarchy of the Forest Service. From time to time, individuals approach the functionaries of the political parties for help (but it appears the functionaries then contact Members of Congress because party officers carry less weight than Congressmen with Forest Service field personnel). By one means or another, then, citizens seeking to obtain or overturn a ruling by a Ranger fight to get their way, and they sometimes succeed.

Every Ranger and former Ranger interviewed in the course of this study has been involved in appeals cases of some kind. It is accepted as one of the hardships of doing public business in a democratic government and is not ordinarily treated as a discredit, even if a field officer is eventually overruled. Yet it is a bother, at best—a distraction from the more "productive" labors of the members of the Forest Service, a cause of additional paperwork, a generator of inspections and inquiries from higher levels. And it is certainly true that, at worst, a Ranger whose constituency is

[16] See pp. 153-155.

constantly restive and rebellious is likely to stimulate some doubts about his judiciousness and skill. So Rangers prefer to avoid them if they can, and are often confronted with a delicate choice between the annoyance and risk of continuous skirmishing with local interests on the one hand, and conceding away elements of the Forest Service program (perhaps to save the remainder) on the other.

Sometimes they make concessions. When his forest supervisor wanted to spray herbicides on stands of commercially undesirable species of trees (in order to destroy the valueless trees and make way for more merchantable growth), one Ranger persuaded him to delay the project indefinitely because the Ranger anticipated that hunting and fishing clubs and other associations of wildlife enthusiasts would raise a hue and cry about the alleged injury to birds, small game, and fish. In another case, a plan to require grazing permittees to put ear tags on their cattle because the Ranger discovered they were running more animals than their permits allowed was deferred when the permittees organized resistance to the program. Another Ranger elected not to press for the termination of a special-use permit under which a small town within the boundaries of his district used national forest property for a town dump; although the dump was an eyesore and a potential fire hazard, he thought the town officials could muster enough support to defeat such a move. On one district, the Ranger was trying to eliminate grazing from some high-altitude, snow-covered, water storage areas, but only slowly and cautiously, so as to minimize the opposition this would provoke. On another, protests by nature lovers concerned about songbirds and game forced deferral of spraying designed to eliminate the highly destructive spruce budworm.

In every instance, to have pressed forward regardless of local sentiment unquestionably would have cost far more in the long run than was gained in the short run; tactically, the concessions were certainly sound. But these examples do indicate how local pressure can influence the behavior of field officers, slowing or modifying what actually happens on the ground as compared with what is mandated from central headquarters.

Thus, a needed step is not taken here, an undesirable permit is issued there, a measure is recommended in order to avoid trouble—and an agency-wide policy can be eroded. From the standpoint of this study, it does not matter whether acquiescence by field officers in the demands of local special interests results from the assimilation of Service personnel into the communities in which they live or from the overwhelming nature of the pressure brought to bear upon them; the erosive impact on policy is the same. Nor does it matter that no single action, as the illustrations demonstrate, is likely by itself to have much effect on policy; multiplied many times, over long periods, in large numbers of Ranger districts, the cumulative impact could be considerable. Unity does not demand uniformity, but it does require consistency and co-ordination. It is in this sense that "capture" of the men "on the firing line" is a challenge to the unity of the Forest Service, as it is to the unity of any large organization.[17]

Personal Preferences of Field Officers

Men do not enter organizations devoid of opinions, values, preferences, and their own interpretations of the world. Nor

[17] Even higher officers are vulnerable in this respect. In 1933, the leaders of the Department of Agriculture under a new administration looked over the Forest Service and "thought that the Washington headquarters had fallen into a rut, being content to relay the legislative demands of the district foresters [now called regional foresters] who in turn were caught in complexes of local interests," (A. W. Macmahon and J. D. Millett, *Federal Administrators* [New York: Columbia University Press, 1939], p. 346.) Consequently, there was a fairly sweeping reorganization in both the regions and in Washington. (See *Report of the Chief of the Forest Service, 1936*, pp. 4 ff.)

do they shed all these once they become members. True, these things may be modified by organizational experience. But job experience is only part of a person's total experience; many of the predilections each man brings with him to his work are reinforced elsewhere and therefore persist even when they are not in harmony with the objectives or desires of his organization's leaders. Since personal predilections and prejudices are presumably among the determinants of behavior, they can produce actions that clash with the proclaimed policies of the organization. This possibility is not confined to the field levels of any agency, of course, but it is especially problematical there because the leadership opportunities to manipulate individual outlooks by personal contact are more limited, and because so many other factors at the lower levels also generate centrifugal forces.

The top officers of the Forest Service were reminded of this when, in the immediate post-World War II period, they adopted a firm position in favor of public regulation of timber cutting on privately owned lands. This has been a controversial question among foresters for two generations; with three-quarters of the forest acreage of the country held by private owners, it is clear that the public forests alone cannot safeguard the nation's forest resources, but many foresters, while recognizing this, share the ideological opposition of many other citizens to expansion of governmental regulatory powers. Nevertheless, advocacy of such a program became official Forest Service policy, and communiques went out from the Washington office requesting field officers to encourage grassroots support by taking every opportunity to explain it, analyze it, and justify it. Apparently, these activities were conceived as a prelude to a campaign for enactment of the necessary legislation.

There are many reasons why the effort made little headway, but one of them was certainly the indifferent response of many field men to the request that they press the issue.

Some were hostile, most cared little. A number made token moves to satisfy minimum requirements, many did nothing at all. The change of administrations in 1953, the enactment of regulatory laws by a number of states, and the improvement of private industrial practices in response to rising stumpage values of timber diminished the urgency of such a program and might have doomed it in any event, but the attitudes of the men virtually assured its downfall. The personal values of Forest Service personnel helped defeat it.

Similarly, it has taken considerable effort by Forest Service leaders to overcome the objections of some of their subordinates to "controlled burning" as an effective and economical way of encouraging new growth in some kinds of forested areas. According to one school of thought, fires deliberately set and controlled to clear away underbrush speed the germination of seeds and the growth of seedlings; for a long time, another school of thought argued that the heat injures seeds, destroys seedlings, and devastates the soil, and that the risk of fires getting out of control is very great. Even to experiment with the technique, the Forest Service had to engage in a large-scale program of information and education for its own men, to sell them on the utility of at least *testing* it, and to overcome the overwhelming fear and hatred of fire instilled in them in their training. The program could not be instituted at once by fiat; the attitudes of many men had to be changed gradually before it made much progress.

To cite a third instance, when a change of administration brought about a change in land acquisition policy—a contraction of the boundaries of the "purchase units" in which further purchases of property for addition to the national forests takes place, requiring the disposal by exchange of parcels lying beyond the new boundaries—one Ranger, reluctant to see acreages reduced, deliberately concentrated his land exchange program only on plots too small and too isolated to manage effectively, and made little effort to dispose of larger holdings located outside the diminished territory. Since he did not subscribe to the philosophy of diminishing the number and scale of federal business enterprises, and anticipated that the policy would change again one day, he "dragged his feet" on carrying it out.

That is not to say that Rangers run their districts as personal fiefs. But they are not just instruments of the agency leaders, either, without wills of their own. While their wills may often coincide with the desires of the leadership, conflicts also develop, and the result is the attenuation or distortion of facets of announced agency policy. A comprehensive national policy may dissolve into a host of different policies in the actual work in the field if such developments are not corrected or prevented.

The Ideology of Decentralization

The Forest Service has made decentralization its cardinal principle of organization structure, the heart and core of its "administrative philosophy." "The Forest Service," says the *Forest Service Manual*, "is dedicated to the principle that resource management begins—and belongs—on the ground. It is logical, therefore, that the ranger district constitutes the backbone of the organization." A former assistant chief described "The constant effort to decentralize and delegate authority to the tree and grass roots," and declared, "Most of the responsibility for national forest work is delegated down to the forest supervisors and the forest rangers." The Ranger, he added, "is checked closely against policies and regulations and must conform, but because it is a fundamental national policy that the forest take its place locally as a contributor to community prosperity, the Chief of the Forest Service insures that the ranger's authority is protected and that no one above him sabotages his planning or action.

In other words, he has his job and is protected in it."[18] Field personnel share this attitude, and speak of Rangers as "their own bosses," men who "run their own show," "kings of their own domains" who "make 90 to 95 per cent of the decisions made on their districts." At higher levels, Forest Service officials often reminisce nostalgically, and rather proudly, of their days in the field. The doctrine is widely accepted—indeed, almost sanctified.

It has a long tradition. It was first propounded in a letter from the Secretary of Agriculture to the Chief when the Forest Service was established in 1905:[19]

> In the management of each reserve [now called national forests] local questions will be decided upon local grounds . . . General principles . . . can be successfully applied only when the administration of each reserve is left largely in the hands of local officers, under the eye of thoroughly trained and competent supervisors.

In the context of the times, such a philosophy was almost inescapable. In the first place, the back areas of the reserves were highly inaccessible, unserved by roads or railroads, unsettled, and without means of communication; a man went into the woods and was virtually isolated for days and weeks on end. One of the earliest United States Forest Rangers was sent out with instructions

> to take horses and ride as far as the Almighty will let you and get control of the forest fire situation on as much of the mountain country as possible. And as to what you should do first, well, just get up there as soon as possible and put them out.[20]

[18] E. W. Loveridge, "The Administration of National Forests," in U. S. Department of Agriculture, *Yearbook of Agriculture, 1949*, pp. 376-377.

[19] Dated February 1, 1905; reproduced in *Forestry Directory* (Washington: American Tree Association, 1943) pp. 108-110. The letter bears all the earmarks of having been drafted by Gifford Pinchot for the Secretary's signature.

[20] B. M. Huey, "The First U. S. Forest Ranger," *Journal of Forestry*, Vol. 45, No. 10 (October, 1947) p. 765.

With little more than their orders and water buckets, rakes, axes, shovels, maps, and badges, almost alone in the forests, the first Rangers went out to execute national policy. There was no real alternative to decentralization.

Moreover, Gifford Pinchot, the first Chief, was determined to avoid the scandals that had discredited the General Land Office, from which the Forest Service took over the administration of the forest reserves. The Division had been beset by fraud, bribery, and laxity;[21] it had also been centralized, authoritarian, and led by men unfamiliar with field conditions:

> The abysmal ignorance of the Washington Office about conditions on the ground was outrageous, pathetic, or comic, whichever you like. Division P ordered one Supervisor to buy a rake for himself and another for his Ranger and rake up the dead wood on the Washington Forest Reserve—a front yard of a mere three and a half million acres, where the fallen trees were often longer than a city block and too thick for a man to see over. And that was no case by itself. Another Supervisor got a similar order for the Lewis and Clark Reserve. But that was only three million acres.
>
> Said Major F. A. Fenn . . . who became one of the very best forest officers the West has ever produced: 'For an officer in the field to question the policy outlined in the regulations on any point no matter how trivial, or to suggest that a change in regulation would be beneficial to the [Government] Service or conducive to better administration, was almost equivalent to religious heresy.' And when out of his practical experience Major Fenn ventured to suggest certain changes in the regulations, he got this reply: 'It is the duty of forest officers to obey their instructions and not to question them.' [22]

Pinchot therefore set out to prevent these faults by building an organization that was composed of professional foresters,

[21] L. D. White, *The Republican Era: 1869-1901* (New York: The Macmillan Co., 1958), pp. 205-8.

[22] G. Pinchot, *Breaking New Ground* (New York: Harcourt, Brace and Co., 1947), p. 162.

co-operative as well as hierarchical, and, inevitably, decentralized.

So decentralization became one of the commandments of the Forest Service from the very start. The idea has been affirmed and reaffirmed, over and over again, for every generation of foresters. It is now part of the dogma of the agency.

Men indoctrinated with this philosophy, it seems reasonable to infer, will probably display a good deal more independence of judgment and readiness to challenge higher authorities than those trained in an ours-not-to-reason-why environment or in the Prussian kind of total military submissiveness that allegedly led Frederick the Great to remark his system was still not perfect because it did not control his soldiers' breathing. To be sure, a Ranger district could not be successfully administered by such docile individuals; performance of the job demands a relatively high degree of personal autonomy. At the same time, the emphasis on decentralization places a high value on assertion of independence, willingness to make decisions and to act without consulting superior officers, and defense of personal and local points of view. It generates centrifugal drives. It builds up the sense of local command. It proclaims that a high degree of self-containment is prized. Thus, it adds to the tendencies toward fragmentation.

Conclusion: The Impulse toward Disintegration

Were the Forest Service not as large and complex as it is, had its massive responsibilities not been imposed on it all at once, did it not depend so heavily on its officers in the field to make the kinds of decisions that translate its enunciations of policy into tangible actions and accomplishments, it would still have to contend with powerful forces drawing the

Rangers along many different paths. For there are many influences besides the top policy pronouncements shaping their behavior. The customs and standards of the groups they work with, the values and attitudes and pressures of the communities in which they reside, and the preferences and prejudices they bring with them from their extra-organizational experiences and associations may lead them in a variety of directions. And the problems of internal communication make the task of directing them a complicated and difficult operation, leaving them vulnerable to the fragmentative influences.

Unchecked, these influences could produce such diverse Ranger district programs as to dissolve the Forest Service into an aggregate of separate entities, destroying it as an integrated, functioning organization.

But they are not unchecked. They are overcome, or at least neutralized. The remainder of this volume is an analysis of the techniques by which integration is achieved.

[illegible]

PART TWO

TECHNIQUES OF INTEGRATION

CHAPTER IV

PROCEDURAL DEVICES FOR PREFORMING DECISIONS

Since it is clear that the organizations for national forest administration might disintegrate if each field officer made entirely independent decisions about the handling of his district, many decisions are made for them in advance of specific situations requiring choice (once experience has indicated the kinds of situations likely to develop). That is, events and conditions in the field are anticipated as fully as possible, and courses of action to be taken for designated categories of such events and conditions are described. The field officers then need determine only into what category a particular circumstance falls; once this determination is made, he then simply follows the series of steps applicable to that category.[1] Within each category, therefore, the decisions are "preformed."[2]

[1] Much social science literature dealing with the "impersonality" of the bureaucracy, its emphasis on proper procedure, and its "rule-orientation," as well as everyday complaints about "red tape," reflect the practice in large-scale organizations of assigning every event or "case" to a category and rigidly adhering to the course of action prescribed for that category.

[2] For the concept of the preformed decision, I am indebted to Dr. Melvin Thorner, of the Medical School of the University of Pennsylvania.

Authorization, Direction, and Prohibition

The description of the course of action applicable to any category may be permissive. It may spell out several series of steps among which the employee shall choose. It may allow the option of acting or not acting, but define the steps (or alternative series of steps) to be followed if the decision is to act. A description of this kind will be called an "authorization" in this volume.

An authorization "permits" an action by guaranteeing the person who takes it that no one in the organization will impose sanctions, or cause (or seek to cause) sanctions to be imposed, on him for so doing. Indeed, in practice, it goes further; an authorization generally turns out to mean the resources of the organization will be used to defend an individual attacked for acting in pursuance of the authorization. The effectiveness of an authorization—i.e., the extent to which it results in the behavior desired by those who issue it—thus depends ultimately on the capacity of the authorizing principal to enforce his guarantees to his agents. If any authorizing principal finds himself unable to do so, he will normally invoke the support of the officers who authorized *him* to act, and the problem may be passed along the line until it reaches an individual or group of individuals capable of sustaining the guarantee. Both public and private organizations appeal to courts, legislatures, and chief executives (or even to electorates) as last resorts.

An authorization is in this sense a "grant" of power. By eliminating, or at least reducing, the risk of personal hardship, it frees the agent to do the specified thing in the specified way. Obviously, where there is no likelihood of attack from any quarter, authorizations are unnecessary; most human behavior needs no authorization. When challenge is probable, or even possible, however, authorizations are normally issued, and individuals often expend great quantities of energy in trying to get more secure guarantees from stronger guarantors.

But an authorization is a limit on behavior, too. It serves notice on those to whom it is issued that if they handle a situation in a designated category in any fashion other than those specified, they do so at their peril. Members of their own organization may proceed against them, formally or informally, or may encourage others (e.g., the Department of Justice or injured parties) to institute proceedings; in any event, even if members of the organization do not initiate or support challenges, they will not defend the violator against attack from other quarters. This is the danger of acting *ultra vires*—beyond one's authorization. Since authorizations are not normally issued, whether in writing or orally, unless the possibility of attack is believed to exist, they advise their recipients what they may *not* do as well as what they may do.

Direction is even more confining. It resembles authorization in that it describes courses of action to be taken by designated individuals should events and conditions in specified categories occur. But directions ordinarily leave no options to act or refrain from acting; they constitute notice that if cases of a given class arise, failure to take the prescribed steps will result in the imposition of penalties. They are descriptions of what must be done in particular circumstances.[3]

Authorization and direction promote and channel action. They are supplemented by prohibitions, which are promulgated to prevent designated actions by establishing penalties

[3] A special form of direction is target-setting. Targets, or "objectives," are directives containing future (sometimes indefinite) dates on which specified quantitative and qualitative levels of output and performance are to be attained. As conditions to be approached, they are somewhat more ambiguous than directives that apply at once, but, as conditions towards which performance should be progressing over time, they constitute an important kind of directive and have significant effects on behavior.

for those who commit them. That is, the limits of formal powers are not left completely to inference from the terms of each permission or instruction; they are often made explicit.

Although authorizations, directions, and prohibitions (and, indeed, goals) may accurately be described in the formal sense in terms of penalties and immunities from punishment, it is quite clear that they do not depend for their effect entirely, or even mostly, on fear of organizational sanctions.[4] Far more importantly, their effectiveness turns on the desire of organization members to observe official requirements, on the feelings of guilt—the pangs of conscience, or, in a manner of speaking, the intrapsychic sanctions—aroused in members who violate official requirements, or on the neutrality of members' sentiments with respect to particular requirements; many of these attitudes, it will be seen later,[5] are deliberately established by the leaders of the organization. In every conversation with field men in the Forest Service, it quickly becomes evident that anxieties about sanctions are by no means absent; it also becomes apparent, however, that other factors play a major part in producing adherence to requirements.

By issuing authorizations, directions, and prohibitions, it is therefore possible to influence the behavior of the members of organizations.[6] An extensive, elaborate network of such issuances envelopes every district Ranger. The network is anchored in more than eighty Federal statutes providing explicitly for the establishment, protection, and management of the national forests; in scores of Presidential proclama-

[4] For a general discussion of the reasons why men obey instructions, see H. A. Simon, D. W. Smithburg, and V. A. Thompson, *Public Administration* (New York: Alfred A. Knopf, 1950), pp. 188-201. See also, C. A. Barnard, *The Functions of the Executive* (Cambridge: Harvard University Press, 1948), pp. 168-69.

[5] Chapter VI, below.

[6] However, the obedience to such issuances is never complete, for reasons discussed in the preceding chapters.

tions and executive orders on the same subject; in hundreds of rules, regulations, and orders of the Secretary of Agriculture; in many court decisions. It is also rooted in uncounted statutes, Presidential orders, departmental rulings, and regulations of staff agencies (the Civil Service Commission, the Bureau of the Budget, the General Services Administration, and others) governing the federal service over-all. But it is not to them directly that the Rangers look to find out what they are authorized, directed, and forbidden to do; for the Rangers, the "bible" is the *Forest Service Manual* put out by the Washington office of the Forest Service, which incorporates, explicates, and interprets the relevant legal documents applicable to the agency, and which contains also additional provisions promulgated by the Washington office under the authorizations in those documents.

The *Manual* currently in force consists of seven volumes. Three more were projected to complete the series, but, before the job could be finished, complaints about its unwieldiness led to a revision and simplification now in progress. Until this is done, however, these volumes remain in effect, serving as the agency Baedeker.

Four of the seven volumes—those dealing with General Administration, Fiscal Control, National Forest Protection and Management, and Acquisition of Lands—are issued to Rangers. (The others are concerned with activities in which Rangers have little or no part, to wit, State and Private Forestry, Forest Research, and Administrative Statistics.) They run to more than 3,000 pages, and it is difficult to think of anything likely to happen on a Ranger district that will not fall fairly unequivocally into one or another of the hundreds of categories catalogued in this *Manual*; indeed, only a fraction of the *Manual* covers most of the recurrent problems of the average district, the remaining provisions applying to events that are not ordinary occurrences anywhere, but which may conceivably come up, or may in fact have already developed here or there. The provisions describe what is to be done, who is to do it, how (and how well) it should be performed, when (or in what sequence) each step should be taken, where the action should take place, and even explain the "why" of the policies—the reasons for their adoption, the objectives they are expected to attain.

The volumes of the *Manual* are looseleaf binders. Additions are inserted at appropriate points; rescinded portions are removed; amended portions are inserted after the changes in the original sections have been posted. In the course of a year, hundreds of additions, rescissions, and modifications are issued from Washington; just getting them filed and posted takes many hours every month. But, in this fashion, the categories of authorization, direction, and prohibition, are constantly defined, made more precise, and kept up to date as errors, omissions, uncertainties, and conflicts are corrected.

Thus, many Ranger-level decisions are made for the Rangers. From free-use permits to huge sales of timber, from burning permits to fighting large fires, from requisitioning office supplies to maintaining discipline, classes of situations and patterns of response are detailed in the *Manual.* Every action is guided. Since different configurations of problems arise on different districts, there must be variety in the Forest Service, but since there is a standard way of handling each class of problems, there is relative uniformity in the way each problem is treated regardless of where it occurs. This is one of the reasons it is possible to transfer professional men in the Service freely, as described in a later chapter.

Naturally, there is more to being a Ranger than just knowing the *Manual*; that is why there are so many additional devices for preforming decisions and otherwise influencing Ranger behavior. Nevertheless, it would be impossible to be a Ranger *without* having a working knowledge of the *Manual,* for it is here that the things forest officers may do, must do, and must not do are in the first instance defined.

Each region, in addition, puts out its own authorizations, directions, and prohibitions controlling field personnel. They take the form of supplements to the Service-wide *Manual*, interpreting and clarifying and rendering more specific the materials emanating from Washington so as to fit them to the needs of each Region. Printed on paper of a different color, but using the same system of classification, they are inserted in the volumes of the *Manual* beside the sections to which they refer; like the Washington office, the regional offices issue additions, changes, and rescissions, and scarcely a day goes by without at least one arriving in each Ranger's mail. Service-wide regulations form a fine mesh governing field decisions, but the mesh is tightened considerably by regional specifications.

Some forest supervisors also issue specifications for insertion in the *Manual* next to regional regulations and bearing the same relationship to regional supplements that regional supplements do to Forest Service regulations. At one time, these were printed on paper of a third color and placed in the proper volumes; recently, however, it has become more customary for supervisors to have their own instructions filed in separate binders for ready reference. They are also in looseleaf form, and, while far fewer in quantity than regional and Washington rules, they fill in whatever gaps may still remain.

Over and above these administrative manuals are technical handbooks describing minutely the conduct of technical operations. Some are published by the Washington office, most by the regional offices.[7] They set forth in detail the standards

[7] The new *Manual* currently in process includes handbooks for most functions prepared in the Washington office. The regional offices will, for the most part, merely supplement some of these. Despite the change in the sources of the technical handbooks, however, the purposes and uses of these documents will remain essentially unchanged.

and procedures for timber surveys and valuation; construction and maintenance of recreation areas; location and construction and maintenance of roads; automotive and equipment maintenance; design and procurement and erection of signs; siting and building permanent improvements (warehouses, lookout towers, etc.); planting trees; fire reporting and damage appraisal. In different regions, depending on the character of their workloads, one finds different books, but none of the Ranger districts visited in the course of this study had fewer than a half dozen on hand. They add hundreds of pages of instruction for field personnel.

Some regions issue "Guides" for field personnel. These pull together the essence of existing regulations and assemble them, with explanations and additional requirements, in handbooks that are somewhat easier to read and follow and consult than the formal rules. In one region, for example, there is a "Personnel Guide," setting forth, step by step, the process of hiring people, getting them paid, keeping files on them, etc.; here, the Rangers find listed every form required for every kind of personnel action, and instructions on the execution of the forms. The same region puts out a "Procurement Guide," explaining how to obtain needed and authorized supplies and equipment, and a "Property Guide," covering methods of obtaining and accounting for government property employed on the job. Indeed, this region even has a "Procedural Guide for Fund Raising Campaigns such as Community Chest, Red Cross, National Health Agencies, etc.," which describes what is permitted and forbidden in solicitation of contributions from Forest Service personnel for these purposes. To be sure, these make life easier for the Ranger, but they also confine him to predetermined courses of action. Practically nothing is overlooked.

Finally, when most of the functions that make up resource management attain a level of activity higher than can be handled by cursory, rule-of-thumb methods, formal district plans for them are drawn up. Indeed, for two functions, the Washington office requires every Ranger district to have a plan; there is none without a fire plan and a timber plan.[8] For the others, Regional offices establish requirements. Occasionally, a supervisor may insist that his Rangers prepare plans for functions not covered by plans on other forests in the Region. Sometimes only one or two districts in a region or a forest may be deemed by regional or forest officials to have a volume of business of a given type to warrant the formulation of an explicit plan. Consequently, except for timber and fire, individual Ranger districts differ in varying degrees with respect to the number and kinds of plans applicable to them. Some districts have as many as half a dozen or more (including, in addition to the mandatory timber and fire plans, land use, watershed, wildlife, recreation, grazing, transportation, planting, and others).

Plans, at least as they are treated in the Forest Service, are preformed decisions. They set long-range (eighty to a hundred or more years for a function like timber management; five, ten, or twenty years for others) quantitative and qualitative goals, break these down into shorter-range objectives, and sometimes reduce these to annual targets. They spell out the steps and stages by which the goals are to be achieved, including methods of operation, and priorities by geographical area, in each district. Out of these functional plans grow the substantive targets and quotas of the Service as a whole. At the same time, once adopted, the field plans govern the actions of the field officers and their work crews; if they depart from the procedures, or fail to fulfill their quotas, and the departures are detected, they may be called to account just as if they had violated authorizations or directions or prohibitions in the *Forest Service Manual.*

[8] However, one Forest Service commentator pointed out that, "On a . . . district which has very little current sales business, the Timber Management Plan generally becomes merely a 'Policy Statement' which sets forth certain objectives very briefly."

Of course, Rangers themselves play major roles in drawing up functional plans relating to their districts. They do most of the field work—the data gathering, the studying, the assessment of needs and opportunities, the evaluation of methods, the forecasting, the strategies and tactics of management. Hence, it might appear that the decisions contained in these documents are their own rather than preformed issuances from above. But the format of their plans is normally suggested by guidebooks and handbooks, if not by the *Manual* and its supplements and by individual instructions from higher headquarters. And staff assistants from forest headquarters commonly participate extensively in every phase of plan formulation—checking field work, reviewing calculations and estimates, offering and vetoing ideas. Since plans must be approved at least by the forest supervisors, often by regional foresters as well, and (in the case of timber management) even by the Chief, forest supervisors generally review carefully and revise, or direct revisions, on the basis of their own studies, while representatives of regional offices check the soundness of plans in the course of regular inspections or sometimes by making quick special surveys of the field situation themselves. In spite of the part the Rangers play in planning, then, the plans may be considered a form of directive embodying preformed decisions.

Timber management plans furnish an illustration. There is one for each "working circle" in the national forests. (A working circle is an area that, by virtue of the character of the timber market and the trees it contains and the terrain and the communities lying within it, is best managed as a single unit; generally speaking, working circles and Ranger districts are coterminous, but not necessarily, and some working circles comprise two or more districts.) The timber plan for every working circle is approved by the forest supervisor of the forest concerned, the regional forester, and the Chief of the Forest Service, and it is reviewed (and bears the initials of) functional specialists for all the related functions (water, recreation, wildlife, etc.) at every level.

The plans describe the historical, economic, social, topographical, and silvicultural features of the territories, the details of timber stands and growth and accessibility, and the ideal pattern of cutting and selling to realize the full potential of the properties. They describe the compartments to be cut, the amount of cutting by acreage and volume of wood, thinning practices and preferred densities of various stands, methods of identifying mature trees to be left standing to seed cut areas, proper marking rules (no tree may be cut in the national forests unless it has been marked or otherwise designated for cutting by Forest Service personnel), and rules for measuring volume (including detailed formulae developed for statistical sampling stands when stands are sufficiently homogeneous in species and age groups to permit measurement by sampling). They often make different provisions for each species. And, above all, they state cutting goals in volumes of timber (both pulpwood and saw timber) and acres, usually for at least ten years, and establish moving three- to five-year plans and annual plans for cutting. So authorizations and directions and prohibitions are not confined to control of paperwork or administrative relationships within the Forest Service, or between the Forest Service and the public; they extend out into the field, into the woods, and govern the physical handling of resources.

All functional plans of this formal kind, combined with the guides and handbooks and the *Manual* with all its supplements, constitute an impressive network of standing orders influencing Ranger behavior. They are not the whole network, though. For there is a steady flow of *ad hoc* instructions from higher headquarters to the Ranger districts—memoranda, letters, circulars. And there are inspectors (described later) and visitors from above who issue informal, oral directives in the course of their sojourns in the field.

Intermittent, irregular, unpredictable, these are usually directed to very limited aspects of district management, and are of temporary duration. All the same, in the aggregate, added to the other types of preformed decisions, they provide the finishing touches to a remarkably complete means of administrative control touching every facet of official Ranger activity.

Clearance and Dispute Settlement

Yet authorization, direction, and prohibition are only *one* category of preformed decisions. Equally important in the day-to-day functioning of a district is the process of channeling decisions proposed by Rangers through higher headquarters before permitting them to take effect—that is to say, before investing them with the immunities and guarantees implicit in formal authorizations. This enables supervisors and regional foresters and their respective staffs to reshape such proposed decisions, and thereby to determine in advance what will actually take place on the Ranger districts.

The formal mechanism for ensuring review is limitation of authorization. Much of the business on a Ranger district involves transactions that can be legally completed only by higher headquarters; a sale of timber worth more than two thousand dollars, for example, can legitimately be consummated only by a forest supervisor (or by a regional forester if the volume exceeds 10 million board feet, or by the Chief if it is 50 million board feet or more), and only very small sales are below these limits. Similar restrictions apply to other functions. Indeed, practically the only actions relating to Service policy that can be completed by the Rangers over their own signatures are, in addition to small timber sales, small trespass cases (up to $100 in value in most regions), issuance of free-use permits (for collection of minor amounts of dead and down timber and similar minor operations), small purchases of expendable items (up to $100 in some regions, up to $500 in others), and the hiring of laborers for timber stand improvement work, for planting, and for temporary duty as lookouts (at prescribed rates and under prescribed conditions after funds have been made available for this purpose), and for fire fighting in emergencies. Virtually everything else is transmitted upward for approval and signature. While the Rangers and their subordinates do most of the physical and paper work of preparing items for higher action, the actions are not binding until the approval is obtained. Sometimes it comes almost automatically; sometimes proposals are radically modified or even rejected. The decision rests with the higher officers.

Clearance is complemented by dispute settlement as a means of bringing policy questions to the attention of higher officials for resolution. From time to time, a supervisor's staff assistant specializing in a particular function (or group of functions on the smaller forests) takes issue with the way a given Ranger manages the function that is the staff man's specialty. Staff assistants concentrating on recreation, for example, are wont to complain that this function is not given due attention, or that some activities charged to recreation management accounts would be more appropriately charged to fire control or something else. Timber management assistants have ideas of their own on how the Rangers ought to manage the timber resources of their districts. Information and education assistants—responsible for the public relations function—tend to urge Rangers to do more in community relations than the busy line officers customarily like to do. In fact, each staff officer at every level, since his energies and attention are concentrated on one segment of the total spectrum of Forest Service policy, displays an inclination to feel more can be done in his function than is actually done by the men in the field. Some of them gradually, and probably inadvertently, edge over from exerting pressure to see that their work is adequately done to commanding line officers as to precisely what ought to be done. (One Ranger complained irritably that his predecessor, who had become the forest supervisor's staff assistant in charge of fire control, seemed to be trying "to show that the district is going to hell since he left.") A few staff men—most of whom were once Rangers themselves—cannot resist the temptation of actually *doing* the work, dealing directly with timber operators, stockmen, and permittees on Ranger districts.

If a Ranger gives in to a staff officer, or if a staff officer does nothing about a Ranger's resistance to his actions or recommendations, such clashes subside. If a staff man attempts to pressure a Ranger into compliance, the Ranger will ordinarily protest to the forest supervisor. If a Ranger objects to staff interference or ignores staff suggestions, the staff officer may carry his case to the forest supervisor. In either event, the supervisor convenes the disputants, hears their arguments, and adjudicates the conflict. Almost without exception, this settles the matter.[9]

The net effect of this procedure is to call to the attention of the forest supervisors (and higher line officers) policy alternatives in the management of Ranger districts that might

[9] Forest supervisors maintain the same relationship to the staff officers in the regional offices as the Rangers do to the staff officers in the supervisors' offices.

Theoretically, a line officer or a staff officer unhappy about the judgment of his immediate line superior could appeal over the head of his "boss." A Ranger, in other words, or a supervisor's staff assistant, might go directly to the regional office. In practice, this is seldom done. To the members of the Forest Service, this would be as intolerable as a captain in the Army circumventing his major and colonel in the line to approach a general. The field officers interviewed in this study were aghast at the suggestion. Moreover, they were convinced that such a violation of the code of ethics would never win them their point, but would surely earn them the disapprobation of all their superiors and colleagues.

Disputes among line officers at the same level are quite rare because they operate in different territories and therefore have few mutual irri-

otherwise go unnoticed. It thus gives them additional opportunities to clear the air of uncertainty, to eliminate ambiguities in standing orders, to say what will be done in particular instances. It suspends the force of decisions until they have been reviewed and approved, modified, or rejected at higher levels. It is a method of preforming decisions in the field that would otherwise not rise for clearance. It employs conflict for purposes of organizational integration.

That is not to say the Forest Service is constantly beset by internal wrangling. Indeed, it is a classic illustration of the process of multiple oversight of administration; although the Rangers, like all line officers below the Chief, are at the focal point of many converging lines of communication from many sources in the administrative levels above them, they find reason to object to only a fraction of the suggestions of the staff men, frequently call upon them for advice and assistance, and manage to work out many differences of opinion without resort to formal adjudicatory proceedings.[10] But the lines of appeal are clear and available to administrative officials, and they are not unused.

The consequences of clearance and dispute settlement, however, cannot be measured by the actual frequency of their employment alone. For almost every Ranger, knowing that works he undertakes and agreements he negotiates and plans he proposes (particularly if these are offensive to one of the agency's clientele) are subject to review and possible change or veto, screens out projects and requests to which the reactions of the reviewers are difficult to anticipate or

tants. They compete with each other for the allocation of funds to their respective territories, but they do not try to tell each other how to manage their areas, which the staff officers over them seem to be doing.

In Civil Service classification, staff assistants are generally equal to line officers at the next lower echelon.

10 Informal achievement of understandings among line and staff officers by such informal procedures has been called "decision by agreement" by Chester A. Shields, Management Analyst Officer for the Rocky Mountain Region, and by others in Region Two.

likely to be negative, and concentrates instead on those more apt to win approval. If a project seems particularly desirable or necessary, or an applicant for the purchase of timber or the use or exchange of national forest property is especially insistent, and a Ranger therefore feels under pressure to proceed along a doubtful line, he normally queries his supervisor or his supervisor's staff assistants before acting. Sometimes, unwilling to risk the embarrassment of having an applicant go over his head and possibly win approval for what he denied, or of commencing negotiations only to be overruled, a Ranger refers the applicants to higher headquarters in the first instance.[11] Thus, over and above what is required by explicit regulations, there is considerable informal clearance. This avoids some clashes with staff assistants that might otherwise arise, and eliminates some rejections and vetoes and criticisms by higher headquarters. But it also gives officers at higher levels additional opportunities to preform decisions about what goes on in Ranger districts.

Of course, the absence of disputes may just as well be evidence that staff officers are failing to influence the Rangers as that the Rangers are fully compliant. So the anticipation of reactions cuts both ways—but more toward Ranger compliance with staff officers' recommendations than toward staff officers' hesitation to offer advice and suggestions. For staff officers and staff assistants are ordinarily in closer and more continual touch with supervisors than are the Rangers, and they share the supervisors' broader territorial perspectives. While the Rangers will not brook what they regard as interference in their administration of their districts, they also recognize that the shared contacts and vantage points of the line and staff officers at the higher level mean those offi-

[11] Of course, it is equally true that staff assistants tend to temper their own suggestions to anticipated Ranger reactions. Officers at subordinate levels thus influence the behavior of those in higher levels as well as being influenced by them.

cers are likely to see many things the same way—and for valid reasons. So the Rangers are not apt to protest vigorously unless the provocation seems to them particularly great.

Then, too, if no disputes arise a supervisor cannot be sure that excessively compliant Rangers or unduly timid staff officers are not permitting the work in the field to proceed further and further from the objectives proclaimed by Forest Service leaders. Anticipation of reactions simplifies the influence of dispute settlement as an influence on Ranger behavior because there *are* occasional reactions, they are resolved at higher levels, and more often than not are resolved in favor of the staff officers.

Clearance and dispute settlement, as a result, reach far beyond what the formal mechanisms *per se* imply. They are for this reason among the major techniques by which Ranger behavior in the field is molded by the organization.

Financial and Workload Planning

Words thus condition the administrative behavior of the Rangers. But there is even greater eloquence and forcefulness in dollars; money talks yet more persuasively than ordinary language. What the Rangers do, and the level of quality at which they do it, are predetermined to a large extent by the way funds are furnished. Forest Service budgeting is inextricably intertwined with a highly developed system of work measurement and planning. The two together make formidable implements for organizational integration by Forest Service leaders.

THE FORM OF A DISTRICT BUDGET

On the Ranger districts surveyed in this study, the smallest annual expenditure is about $50,000, the largest about

$220,000. This is more or less typical of the Service as a whole; most districts fall within this range.[12]

No one on a Ranger district ever handles the cash (except for isolated petty cash accounts). Revenues from sale of timber and permits are sent by the purchasers and permittees directly to the regional offices. Money available for expenditure takes the form of accounts maintained in the regional offices against which payrolls, requisitions of materiel, and petty cash expenses are charged as reported from the Ranger districts. As payrolls, requisitions, and vouchers come into the regional offices from the field, regional fiscal officers issue warrants to Treasury disbursing officers directing them to issue checks to the payees (or to transfer credits from the accounts of the Forest Service to the accounts of other agencies of the federal government when supplies or services are received from them).

The accounts for each district are not gross sums; they are broken down into many specific categories. The major ones are National Forest Protection and Management, Fighting Forest Fires, Insect and Disease Control, Forest Roads and Trails, Co-operative Work Funds, and the Working Capital Fund.

The National Forest Protection and Management account finances virtually all of the routine activities on Ranger dis-

[12] Revenues on three of the five districts exceed expenditures, in one case by a ratio of five to one. None of the income, however, remains on the districts. All of it is deposited with the regional offices. The regional offices are authorized to hold deposits for co-operative work (see p. 109) in suspense accounts for expenditure on the work for which it was designated in the districts of origin. The regions also hold 25 per cent of the revenues for return each year to certain county governments (see footnote 5, Chapter II). Finally, they set aside 10 per cent of the receipts of the national forests to comply with the statutory mandate to spend 10 per cent of this income on forest roads in the states in which the income originates. All the rest of the money is turned over to the Washington office, which presents a quarterly check to the U. S. Treasury; this year, the cumulative total of such payments reached the billion-dollar mark.

tricts. Fighting Forest Fires is a separate account because the costs of forest fire fighting cannot be accurately predicted; indeed, since they depend so heavily on chance natural factors, this account is normally completed by supplemental appropriations after the actual costs have been determined. Insect and Disease Control is a separate account because it, too, is a less routine function than other phases of protection and management. The Forest Roads and Trails account covers the construction and maintenance of those transportation facilities needed to protect and administer the resources of the national forests. Co-operative Work is an account for sums deposited by purchasers, permittees, and co-operating units of government to pay the Forest Service to do work required by the terms of their agreements with the Forest Service, work (such as betterment of timber sale areas, improvement of ranges, road maintenance and construction, slash disposal, etc.) that the co-operators would otherwise have to do themselves or hire contractors to perform; Congress has authorized the Forest Service to serve, in a manner of speaking, as contractor, accepting payments (which are not covered) into the Treasury to pay expenses, and refunding whatever is not expended on the specific areas involved.[13] The Working Capital Fund is a revolving fund for the replacement of obsolete and worn-out equipment.[14]

[13] Some activities of the Forest Service are thus financed out of several separate accounts. Timber stand improvement, for example, is supported by both Protection and Management appropriations and Co-operative Work funds. Wildlife and range improvements are similarly sustained. Pest and disease control get some money from Protection and Management funds and some from Insect and Disease Control. There are technical differences in the precise kinds of operations each account is used for, but, in the broad sense, any given function may well depend on more than one source of financing, including sources that do not require annual appropriations.

[14] As equipment is used, depreciation charges are calculated and charged against the categories of activity—i.e., the functions—on which it is employed. These sums are then transferred from the functional accounts to the working capital fund. Eventually, worn-out machines

Whether or not Insect and Disease Control appears in the budget of a given Ranger district depends on the acuteness of the infestation or infection; generally, ordinary Protection and Management funds suffice to keep this under control. And whether or not Working Capital Fund appears in a district budget depends on the bookkeeping practices of the region in which it lies; some Regions manage their equipment and equipment budgets in a more centralized fashion than others. The other major accounts, however—National Forest Protection and Management, Fighting Forest Fires, Forest Roads and Trails, and Co-operative Work Funds—show up on virtually every district budget.

The major accounts are divided into sub-accounts by administrative action of the Forest Service. Protection and Management, for example, has twelve subdivisions: Unit Management (i.e., general administration);[15] Timber Use; Insect, Rodent, and Disease Control; Range Use; Land Use; Recreation Use; Wildlife Use; Watershed Management; Fire Control; Reforestation and Revegetation; Maintenance of Improvements; Construction of Improvements. Most of these are further subdivided. Forest Roads and Trails is divided into sub-accounts for construction, maintenance, and other operations, and comes to the districts allocated to specific projects. There are co-operative sub-accounts for timber stand improvement, road maintenance and improvement, range improvements, and wildlife management projects.

and implements are sold to the highest bidders, and these returns are also credited to the working capital fund. Thus, the fund is kept at a level that enables the Forest Service to acquire what it needs as the needs arise.

[15] Rangers are paid out of Unit Management funds from the Protection and Management account. They do not receive money charged to specific functions, and may therefore allocate their energies as they see fit. In point of fact, however, they are more or less compelled to divide their energies among the various functions in proportion to the funds made available for those functions since they must supervise the performance of those activities.

Other major accounts may be similarly broken down, though into fewer categories, both by general activity and specific project.

Every cent available for expenditure on a Ranger district is in one or another of these accounts and sub-accounts. The money in each major account must be spent on that activity only. There is greater flexibility with respect to sub-accounts, for limited transfers (not to exceed 10 per cent except with the approval of higher headquarters) from one sub-account to another are permitted, but these are watched carefully by the higher levels. Every payroll, every requisition, every voucher is charged against specific sub-accounts; with supervised and rather restricted variations, manpower, supplies, and equipment are employed on the specific functions for which they are designated.

The Ranger district budget document is thus a table of vertical columns and horizontal rows. The columns are the accounts and sub-accounts. The rows show individual employees, equipment and materials, and individual construction projects.[16] Tracing a column shows how the amount allocated to each function will be spent on the time of various employees, on equipment and supplies, and on construction projects. Tracing a row shows how much of each individual's salary, of each construction project, and of equipment and supplies usage, will be charged against each function. The amount of time each individual in the Ranger's organization can spend on each activity is thus determined by the budget; so, too, is the way equipment use will be distributed among the various functions; so, finally, is the distribution of construction among the various functions (i.e., money for construction of a recreation area cannot be shifted to building

[16] In some regions, these are three separate documents; in others, they are all assembled in one master plan. Where Employment Plans and Equipment Operation and Repair Plans are used, they usually give greater detail than the more general Financial Plan.

a wildlife refuge or a road). What will happen on a Ranger district is controlled by the allotment of funds; it is determined by a pattern of performance budgeting that sets the deployment and employment of the Ranger's work force and materiel over the year. It is an elaborate series of preformed decisions.

WHERE THE BUDGET COMES FROM

In most administrative agencies in the federal government, agency budget requests are assembled each year by putting together estimates from every unit, which are reviewed, adjusted, and consolidated at higher administrative levels; the higher levels then submit to their superiors their own estimates based on their revisions and compilations of the units under them; and the process continues until the estimates of the agencies as entities are completed. From unit to section to division to bureau to department, the figures are passed and reviewed and screened and at last incorporated in a single set of schedules. Finally, they are reviewed by the Bureau of the Budget, approved by the President, and presented to Congress in the budget document and the budget message. This is the way the President's budget is fashioned.

The process begins in the summer of each year. By fall, when hearings before the Bureau of the Budget begin, the departments and their constituent bureaus have compiled their figures and are ready to defend them. By the end of the year, the President's budget is almost completed. By the time Congress convenes in January, it is ready for submittal. All through the spring, Congress holds hearings and enacts appropriations. When the new fiscal year begins on the first of July, appropriations are available for expenditure. Then, almost at once, the cycle starts again.

Like all other bureaus, the Forest Service presents its esti-

FIGURE 7. *An Illustrative District Fiscal Plan**

Name of employee or object of expenditure	Position title, grade and salary	Rate per hour	Protection and ma								
			020	031	033	051	061	063	064	070	08
Clarence E. A.	District Ranger GS-11, $6605 (1)	3.45	(Hours) ($)								
William J. H.	Forester GS-11, $4930	2.55	(Hours) ($)								
J. R. H.	Forestry Aid GS-5, $4210	2.19	(Hours) ($)								
Joe B. P.	Forestry Aid GS-4, $4010	2.04	(etc.) ($)								
James A. O.	Forestry Aid GS-4, $3500	1.82									
John T. F.	Const. & Maint. Sup. CPC-7, $4472	2.28									
W. A. L.	C & M Sup. Asst. $4077 (2)	2.08									
Ben W. S.	Equip. Op. II $3702	1.89									
Floyd G. B.	Sr. Auto. Mech. $4056	2.07									
Laborers		1.13									
Engineer aid		1.65									
James M. W.	Fire Control Aid GS-2 (3)	1.67									
Sarah D.	F. C. Aid GS-2	1.46									
William B. R.	F. C. Aid GS-2	1.59									
Equipment use and materials			.... ($ only)								
Construction			.... ($ only)								
Miller Bridge Rd.	3.0 miles										
Cedar Spring Rd.	2.3 miles										
Cunningham Rd.	2.0 miles										
West Rd.	2.3 miles										
Lick Fork Creek bridge											
Water system											
Bradley gas tank											
Total hours .... Sub-total dollars											

102	110	Total P & M	Roads and trails 120	130	Total R&T	Cooperative work funds 033	034	133	Total CWF	Fighting forest fires—102	TOTAL
											($ only)
											($ only)
					($ only)						($ only)
					($ only)						($ only)
					($ only)						($ only)
					($ only)						($ only)
					($ only)						($ only)
	($ only)										($ only)
	($ only)										($ only)

NOTES

* Simplified. Figures omitted to facilitate reading. Details vary from place to place.

(1) Civil Service Title, Classification and Annual Salary.
(2) Absence of Classification indicates hourly wage rate employee.
(3) Absence of Annual Salary indicates salary paid only when employee is actually on the

Key to Accounts (From new *Manual.* May differ slightly from discussion in text of volume.

020 Ranger District Management
030 Timber Resource Management
- 031 General Timber Resource Management
- 032 Timber Resource Management Plans
- 033 Reforestation
- 034 Timber Stand Improvement
- 035 Nurseries

040 Forest Pest Control
- 041 General Forest Pest Control
- 042 White Pine Blister Rust Control

050 Range Resource Management
- 051 General Range Resource Management
- 052 Range Revegetation

060 Land Management, Minerals
- 061 Special Land Uses
- 062 Minerals
- 063 General Surveys, Maps & Boundary Posting
- 064 Land Exchange, Acquisition, and Disposal

070 Recreation—Public Use
080 Wildlife Habitat Management
090 Soil and Water Management
- 091 Water Management
- 092 Soil Surveys

100 Forest Fire Protection
- 101 Forest Fire Prevention & Presuppression
- 102 Forest Fire Suppression

110 Structural Improvements
120 Maintenance of Roads (Forest Development System)
130 Construction of Roads (Forest Development System)
- 131 Timber-Access Roads
- 132 All-Purpose Roads
- 133 Timber-Purchase Roads

140 Maintenance of Trails (Forest Development System)
150 Construction of Trails (Forest Development System)

* * *

710 Forest Tree Insects & Diseases—Detection & Appraisals } May involve National Forest
720 White Pine Blister Rust Control—Technical Director } and Private Lands

FIGURE 8. *An Illustrative Equipment Use Budget and Financial Plan, Fiscal Year 19—*

Equipment size and class	No. of units	Estimated use		Total use rate (cost per mile or hour)	Total budgeted cost all funds ($)	Distribution of use by functions, funds and/or projects (miles)																	
		Inclusive dates if seasonal	Miles or hours			Protection and management									Forest roads and trails			Cooperative work funds			Fighting forest fires	Pest control	Working capital fund
						020	031	033	034	061	063	070	101	110	120	130	140	033	034	080	102	041	
Jeep, ¼ ton	1	Year	10833	.12	1300		2000					833			1000				7000				
Pickup, ½ ton	4	"	46943	.09	4226	7778	17264			778	400	3318	4422	777	4000	3118	1111			2311		555	1111
Pickup, ¾ ton	1	"	4040	.13	655										500				2000			1540	
Truck, 1½ ton	1	"	2080	.15	150							300		200	500	400				100	580		
Truck, 1 ton	1	"	9806	.105	1030			454	95									545	6762			1950	
Tractor	1	"	50	7.00	350																50		
Fire plow	1	"	50	1.25	63																50		

For *Key to Accounts*, see illustrative district fiscal plan (Figure 7).

mates to its department, then justifies the departmentally approved figures to the Bureau of the Budget, and defends the figures approved by the Bureau and by the President before Congressional appropriations committees. But its estimates are not built up by new requests submitted annually from units in the field. Instead, the Washington office formulates the budget estimates. Field units do not start their fiscal planning until the middle of spring, when there is a fairly clear indication what the Service as a whole will get from Congress; they do not make up their *final* fiscal programs until appropriations have been passed and signed by the President. That is, in the Forest Service, subdivisions of the organization do not explicitly advise their chiefs (as such subdivisions do in many other agencies) how much to ask for by telling them each year how much is needed; rather, they report their estimated volumes of business for the various work programs to Washington, where these are converted to manpower and dollar needs by the Chief's office. In a sense, a Ranger's budget comes down to him from above.

HOW THE FOREST SERVICE BUDGET IS FORMULATED

But only in a sense is this true! Putting together budget requests that can survive the searching scrutiny of departmental budget officers, the Bureau of the Budget, and the appropriations subcommittees of Congress requires demonstration of the relationship between cost and output, between proposed expenditure and performance. This means all fiscal planning rests in the last analysis on information coming up from the field. It is as characteristic of the Forest Service as of bureaus that build their financial structure afresh every year. Despite the fact that the Washington office formulates the requests to Congress without obtaining annual financial estimates from its own field units, it depends on field data to do this task.

The Washington office maintains a standardized list of specific jobs—that is, discrete operations characterized by countable work output at specified standards of quality (perfection and intensity)—that must be done to perform all of the functions in national forest administration. It also keeps on hand records of the "converting factors"—the amount of time (down to the smallest fractions of an hour) needed to perform one unit of work on each job. Finally, it has figures on the number of units of work—the volumes of business—for each job. Multiplying the volumes of business by unit time allowances yields the number of man-hours required to perform all the functions involved in managing the national forests. The man-hours are converted into dollar sums, and expenses for non-personal services (travel, supplies, communication, etc.) are computed as ratios of personal services. Thus, the estimates are formulated in terms of physical performance. The performance information comes in from the field; that is why the Washington office is able to turn out the agency budget without getting dollar estimates from the subdivisions.

It took years of study to identify the component jobs, and to develop the standards of quality and establish the work-unit time allowances for each job, for each function, at each level of organization; there are several hundred of them in the composite job list for Ranger districts (where, of course, the bulk of the physical work is done). The list is revised as experience demonstrates some categories are too diffuse to yield a countable output and others are really parts of broader jobs rather than meaningful jobs in themselves. The Division of Administrative Management in Washington and the Divisions of Operation in the regional offices make adjustments based on field studies from time to time.

Calculating the time allowance for the completion of a single unit of work requires extensive field studies. Using the standard job-list, teams from the regional offices select sample Ranger districts at random and keep careful track of the amount of time expended on each job by Ranger and sub-Ranger personnel in the field, in the office, and in travel, making due allowance for differences in the efficiency, ability, etc., of the men studied. The figures are averaged, taking special conditions into account, and are rechecked once every five years or so; individual items may be rechecked even more frequently if challenged by field personnel. At the Washington level, all areas of the country in which the allowances for each function are relatively uniform are grouped into zones without regard to regional boundaries. For some functions (Land Use Management and Recreation Use Management, for example), the entire country is one zone; for Fire Prevention, on the other hand, there are two zones; for Timber Sales Administration, five; for Fire Suppression, seven. (Different national forests in the same region may therefore fall into different zones for different functions.) Thus, the Chief's office in Washington has the information it needs for budgeting and other managerial objectives—for instance, information adjusted to allow for differences between checking smokes or marking 100 thousand board feet of timber for cutting or scaling (measuring) 100 thousand board feet of logs in the Douglas fir country of the Pacific Northwest (where the holdings of the federal government are consolidated, served by comparatively few roads, and sparsely populated), as contrasted, say, with the Southeast (with its species of smaller trees, scattered government properties, networks of state and local roads, and relatively dense population).

Volumes of business from field units all over the country, reported regularly from Ranger districts to forest supervisors, by forest supervisors to regional offices, and by regional offices to Washington, provide a base for determining financial needs, function by function, for the Forest Service. Consolidated in regional headquarters, assembled into Service plans in Washington, and supplemented by estimates based on Timber Management plans for every working circle, and by inventories of proposals from the field for capital improvement projects, they make it possible for the Forest Service to estimate with a high degree of precision how many units of work it will seek to perform in the coming year; and from this to calculate how many man-hours and man-years of personal service it will have to finance. To convert these into dollars, average hourly wage rates are worked out for every zone and every function. In the case of officers and employees in the classified service, this is relatively simple, for the rates are set for each grade by statutes applicable to the entire government service. For laborers, whose wages are set at prevailing levels by departmental wage boards for each section of the country, these rates are established on the basis of wage studies by field executives; each Ranger, for example, makes an annual survey and reports the results, which are taken into account by the wage boards. Then, the averages are worked out, making due allowance for the fact that some activities demand highly paid workers, some require relatively unskilled labor. With these averages, man-hours for the Service as a whole can be translated into financial terms.

Finally, by studying the amount of travel, supplies, communication, and other non-personal services associated with past expenditures for personnel, fixed ratios are arrived at. Field studies are conducted frequently to determine whether the ratios can be reduced by better planning, and the ratios are changed from time to time. In any given year, however, when the Washington office has computed its manpower costs for administering the national forests, it can tell how much money its men need to do their jobs.

The Forest Service relies on a flow of reports on performance, and of reports on studies of performance, from the field to formulate its budget estimates. In this sense, the field officers participate significantly in the budget process. The fact remains, nevertheless, that the Washington office does not ask them each year how much money they need to operate; it makes its own judgments from workload volumes and work-unit-cost statistics.

APPROPRIATIONS AND ALLOTMENTS

By the end of each calendar year, the Forest Service budget, having been reviewed and adjusted by the Department of Agriculture and the Bureau of the Budget, is included in the President's budget. In January, the budget document goes to Congress, and Congressional hearings on appropriations begin. By the end of spring, before the start of the new fiscal year, most appropriations have been enacted. The Forest Service is then faced with the problem of allocating its appropriations, which are generally somewhat lower than its requests, to its regions; the regions must make their allotments to the national forests; the forest supervisors must allot their funds to the Ranger districts.

These levels prepare for the allotments in advance. During April, regional, forest, and Ranger headquarters start to prepare their own budgets for the fiscal year that starts on the first of July. The Chief generally provides guidelines, assessing the temper of Congress and advising in what functions reductions below the previous year's appropriations may be expected, where increases may be hoped for, and where the line will probably be held. Acting on these guesses and aspirations, tempered by their own predictions of what is likely to happen, the members of the organization start their financial planning.

The regional offices work just the way the Washington office does; they use administrative statistics on volumes of business and averages of unit manpower and related costs to figure out how much they will need. Although allotments

Form 0-47A-R8
March 1957

FOREST SERVICE R-8
PROJECT ANALYSIS SHEET
(Illustrative)

1. Forest S---
2. District L--- C---
3. Project Description Install pump + pump house at Bradley. Pump + storage tanks on hand.
4. Function 110 (Protection + Management) (See Key To Accounts, Fig. 7)
5. Proj. Name Construction- Improvements

6. Superintendence and Labor (Labor need not be by name)

Name of Employee	Superv. (Hrs.)	Labor (Hrs.)	Leave & Hol.(Hrs.)	Adm. Services (Hrs.)	(Per Diem)	Total (Hrs.)	Hourly Rate	Amount $
Forester	8					8	2.28	18.24
Langley		40				40	2.08	83.20
Labor		80				80	1.13	90.40
								191.84

7. Summary:

Superintendence and Labor	$ 191.84
Supplies and Materials	101.56
Equipment Use	6.60
Contract Job	—
Total Direct Project Cost	$ 300.00

Project to Start Aug. (Month) Project to be completed Aug. (Month)

Proposed C.E.A.--- (District Ranger) Approved ______ Funds Allocated ______

Date ______ ______ ______

*These columns are for use when form is used as a functional summary sheet

(Over)

FIGURE 9

8. Equipment Use

Class	Miles or Hours	Rate	Amount
115 (Pickup)	50 mi.	.09	4.50
120 (Truck)	20 mi.	.10½	2.10
			6.60

9. Supplies and Materials (Include Utilities)

Quantity	Item	Unit Cost	Amount
	pipe		26.56
	electrical		45.00
	service		25.00
	lumber + hardware		5.00
			101.56

to regions are not broken down beyond major accounts, regional figures are justified function by function. Thus armed, the regions make their respective claims for their shares of the money actually made available by Congress. Generally speaking, there is less money than they need to do all they would like, and each regional forester tries to get as much as he can for his organization. Objective work-load figures and cost accounting meliorates the bitterness of the competition, but it does not eliminate it by any means.[17]

Even as the regions are exerting their demands upon the Chief, the forest supervisors are making their own claims for their shares in the regional allotments. During April, they work up detailed figures by accounts and sub-accounts, basing them on the same time allowances required to do each unit of work, volume of business figures, and hourly wage rates as are used by the regional offices. These spring plans for the national forests are made up in pencil. In June, after the appropriations to the Washington office have been divided

[17] In an earlier day, the regional foresters assembled like ambassadors at international conferences. Indeed, in one of those jests that embody profound truth, they actually drew up a "Treaty of Washington" that hung for many years in the office of the Assistant Chief for Operation:

> Whereas the race in competitive armaments, in National Forest finance, no less than in wider fields, must in the end make liars out of honest men, and
>
> Whereas the exercise of undue and unrestrained imagination in picturing estimated cost to complete, and necessary cost for adequate must lead to constant bickering among ourselves and with our chief [sic] and
>
> Whereas no intolerable burden should be placed on our imaginations and our bosses' blue pencil, and
>
> Whereas with external enemies hounding our flanks and rear, we must either hang together or hang separately
>
> Now therefore, to the end that these evils may be avoided and that we may live at peace with one another,
>
> We, the undersigned, as lords of our several domains do hereby covenant and agree with each other and all with our Chief,
>
> That hereafter international peace shall be held more precious than allotments, and parity more dear than reactions.
>
> Done in the City of Washington this 5th day of April, 1930, A.D.

up among the regions, the regional offices notify the forest supervisors how much each national forest has been allotted for each function (after conferences in which there is as much bargaining as is possible within the framework of the work load statistics), and the budgets are put in final form and become binding for the fiscal year.

At the same time, the Rangers make their April bids for their shares in the anticipated allotments to their respective national forests. Unlike higher headquarters, they cannot simply employ time allowances per unit of work to their volumes of business in order to determine their needs; these allowances are averages applicable to aggregates of districts, but not any individual one. Rangers therefore compile lists of all the projects they propose to undertake (for example, to cut and sell so much timber; plant so many acres to trees; perform timber stand improvement on so many acres of forested land; clean up camp grounds in accord with a set schedule; etc.), and figure out, on the basis of experience and local conditions, how much labor, supplies and materials, equipment use, and contractual service each project will take. This, in turn, indicates how much each will cost. It is from this information that the tabular charts of proposed expenditures for each district—function by function, employee by employee, machine by machine, project by project—are derived. In the spring, the charts are made out in pencil; then, when the June allotments come through from the supervisors, the charts and lists of projects are revised, put in final form, and the budget is fixed for the new fiscal year. There are sometimes arguments and bargains among Rangers, staff assistants, and supervisors over the final allotments to districts, but, on the whole, by the time the allotments reach this level, there is not much opportunity to have a profound impact on the division. The basic decisions have already been made.

SPENDING

With the start of the fiscal year, in accordance with these approved operating plans, each Ranger begins to issue payrolls, requisitions, and vouchers against his allotments. Every hour devoted by his assistants, aides, and laborers to a given function is charged against the appropriate sub-account. The cost of every mile his equipment runs, every hour it is employed, is deducted from the sub-account for the function for which it is used. Every piece of construction, every mile of road built and maintained, every bit of materials and supplies ordered, reduces the sub-account for some function. If a sub-account (including allowed variances from allotments) is exhausted, no more work may be done on the function it finances unless additional funds are made available by higher level action. Rangers must consequently be careful to spread the work of their subordinates over the appropriate seasons of the year, and to plan the activities of every employee not hired on an annual basis (most laborers are paid only when actually employed) to piece out a full year's work for as many as possible. At the end of the year, the amount actually spent on each function must come reasonably close to the amount allotted for that sub-account, and the totals must fall within gross sums for each major account. The Ranger has control over the timing of his operations (although weather often helps determine this for him), but what has been done on a district when a fiscal year ends must reflect faithfully the figures in the budget.

The Rangers interviewed indicated they would run their districts somewhat differently if they had lump sum allotments against which they could draw for any purpose they wished instead of finely itemized allotments.[18] Most would

[18] "We would like to comment," observed one regional forester after reading this paragraph, "that we in this Region are coming more and more to the thought involved here. . . . We place the major emphasis on

pour more of their energies into timber management, but one would cut down on this to intensify his range management. One would cut down drastically on recreational activities. Another would reduce wildlife management to concentrate on land exchange more heavily. A third would like to see more planting than he is now able to accomplish. One Ranger complained the budget came down each year and "blew up in my face." They would all conduct their operations with at least slightly different emphases. But, for reasons explored in the chapters that follow, they do what finances dictate. The budget, in short, is a major determinant of administrative behavior.

The Problem of Deviation from Preformed Decisions

In the light of the powerful centrifugal tendencies at work in the Forest Service, there is no obvious reason why advance decisions announced by Forest Service leaders in the form of authorizations, directions, prohibitions, clearances, settlements of disputes, and financial allotments should induce Rangers to act in the prescribed fashion. Indeed, it might reasonably be expected that the competing influences on Ranger behavior would prevail over such announcements unless the announced preformed decisions are in some way reinforced.

the job to be done rather than the exact functional allotment of money. We know that if a Ranger District has a sound plan which covers the jobs that should and must be done, then the money will take care of itself and wind up in the right accounting pocket. Thus, we are more and more coming to the point of governing our Ranger funds by an over-all check by appropriations rather than by individual functions. We do still, of course, review the functional budget plan a couple of times a year to be sure we are not getting entirely out of line." Generally, however, as the Rangers see the situation, the accounting controls are still quite detailed and tight despite trends towards increasing flexibility.

They *are* reinforced. Adherence to them is not treated as an automatic consequence of issuance. Deviation is consciously discouraged, conformity deliberately encouraged. The methods of doing so are described in the next chapters.

CHAPTER V

DETECTING AND DISCOURAGING DEVIATION

Reporting

To determine whether behavior of men in the field conforms to the requirements of preformed decisions promulgated by organization leaders, the leaders must obviously keep themselves informed about what actually goes on in the field. The easiest way for them to do so is to ask the field men what they are doing. Hence, reporting is a common characteristic of all large-scale organizations.

Reports are not the only means of finding out what happens in the field. Indeed, if reporting is not supported and supplemented by other information-gathering techniques, it may well become a relatively unreliable method. But it is convenient and effective when coupled with other devices. So it is virtually universal; the greatest part of the upward flow of administrative communications in any agency is likely to consist of reports requested by the agency leadership.

In the Forest Service, requests for reports invariably specify in detail the exact information desired, the format in which it is to be presented, the period it is to cover, the dates on which it is to be filed, and the place to which it must be sent. As a rule, reports are submitted on prepared forms, which thus prescribe precisely what information must be furnished and how it shall be organized.

Most of the reports are statistical summaries covering every activity in national forest administration. They are submitted routinely under standing orders. Each regional office renders more than one hundred different reports each year, most of them annually, but many of them quarterly or monthly. At lower levels, the number of statistical summaries is somewhat smaller, varying from about 65 to 85 for forest supervisors reporting to the several regions, and from two dozen to over 40 for Rangers reporting to their supervisors. The majority are annual, but the field offices, too, must submit a substantial number of more frequent reports. At all levels, the number is large enough to compel the establishment of "promise-card files" and other reminders of due dates; there are too many for anyone to carry all the requirements in his head, and many reports demand ample warning in advance to allow adequate time for preparation.

Since all the work of national forest administration is performed on Ranger districts, and the higher headquarters depend on field men to provide the information on which computations of agency accomplishments are based, it might reasonably be expected that the largest number of statistical reports would originate in the base of the administrative hierarchy, and that the volume would decrease at the higher levels, where information from the field can be consolidated and summarized. Logically, it would seem field officers ought to prepare more reports than their superiors, and the inversion in the Forest Service therefore appears rather curious. There is, however, a simple explanation. In addition to statistical compilations, field men furnish reports on individual actions (such as a sale made, a fire suppressed, a permit issued) from which higher levels can assemble much of the information they need.

Data on individual actions are primarily for financial and

bookkeeping purposes rather than for program and policy control; they permit administrators in the forest and regional offices to maintain surveillance of receipts and expenditures. But they also keep the higher administrative echelons informed about what their subordinates are doing. The payrolls, requisitions, and vouchers sent from the field to be charged against the functional accounts and sub-accounts established for a district constitute a running record of what goes on in the district. By consulting the record from time to time, the Rangers' superiors can keep abreast of the Rangers' activities, and they are able to assemble from such reports the statistics embodied in their own summaries.

Moreover, as noted earlier, many transactions—large timber sales, for example, and special-use permits—must be signed by supervisors and regional foresters before they take effect. As a result, these officers do not have to ask for special reports on many functions in order to compile their own reports; the information is readily at hand.

Routine "administrative procedure," though often treated as something different from reporting, is in effect nothing more than a form of reporting that serves several ends at once. And this, in turn, means higher administrative levels are able to formulate reports on subjects for which no statistical summaries are tabulated by field officers.

Over and above regular, periodic reports of both the tabular and individual-action types, there are frequent calls for special reports on an *ad hoc* basis. In addition, the written documents are supplemented by uncounted informal reports; every time the Rangers get in touch with higher headquarters for guidance or advice or preliminary clearance of a proposed field action or to settle a dispute with a staff man, and every time a visitor from a higher level appears on a district, the Rangers' superiors get new insights into what is happening in the field.

All in all, then, the flow of information from the districts to the forests, the regions, and to Washington is steady, massive, detailed, and comprehensive. In one way or another, the Rangers themselves furnish facts revealing how closely they are adhering to the preformed decisions of the Service leaders, facts that disclose any deviations from the promulgated standards. It is doubtless true, as members of the Forest Service at every level aver, that the system of reporting has not been set up to expose deviations so much as to provide the leaders with the knowledge they need realistically to plan and guide the destinies of the agency.[1] Distrust is not the driving force behind the system. Just the same, whatever the intentions of those who established and maintain the upward flow of reports, one result is to bring to their attention any continued departures from announced behavioral norms.

Theoretically, a field officer who does depart from announced policies, as a result of the tendencies toward fragmentation that pull at all Rangers, could falsify his reports to conceal his digression. In practice, this is seldom feasible, for misrepresentation in one report would soon produce contradictions with so many others that it would require almost all a man's time and energy as well as the most extraordinary ingenuity to tamper with all of them so as to make them consistent. What is more, many people—employees, users of national forest products and facilities—would eventually have to be drawn into the conspiracy. And even if all the reports were successfully altered, the information in the reports would then conflict with that obtained by higher levels through the other channels described in this chapter. It is almost inconceivable that manipulations of the records

[1] A number of reports have been instituted primarily to satisfy Congressional requirements, and are not considered in the Forest Service to be particularly valuable as administrative tools. Yet even reports of this character, though not administratively worth the trouble they take, do add at least a small increment of information to headquarters activities about field activities.

could long escape detection.

In any case, the incentives to falsify reports are not very strong. In the first place, the penalties for occasionally inadequate performance are far less severe than those for misrepresentation: the risks of dishonesty are infinitely greater than those of honesty. Secondly, for reasons explored in Chapter VI, the whole ethos of the Service discourages falsification. The observer of the organization quickly gets the feeling such behavior would be regarded as not only immoral, but cowardly, unmanly, degrading to the individual and to the Service (whose members have a fierce pride in it), and that any man who practices it must end with contempt for himself for not having the courage to fight for those departures from policy that he believes right or to admit his errors when he is wrong.

Consequently, reporting tends to be highly accurate, and field officers turn in information that sometimes reveals their own weaknesses and mistakes as well as their competence and their triumphs. Their superiors therefore have reliable data on which to judge whether actions in the field—the physical operations in the woods and the negotiations with the public—conform to the preformed decisions of the leaders of the Service.

Official Diaries

Rangers, assistant Rangers, and their principal aides are required to keep official diaries throughout the year.[2] The diaries show to the nearest half-hour how each workday is spent. On standard Service-wide forms, the field officers and employees record each thing they do, describing the activity in enough detail for any inspector to identify it, the func-

[2] Generally speaking, officials at higher levels keep diaries only when they are in the field. In the field, in short, *all* forest officers maintain these records.

tions to which the activity is chargeable, the time at which it began and was completed, and the amount of office, travel, and field time it entailed. They thus compile a full running record of the way they employ their time.

A check of the diaries of the personnel on a Ranger district is extremely revealing. Tallying the allocations of time to the various functions of national forest administration, for instance, indicates whether the district organization has actually divided its energies in accord with directives from above. It also indicates whether the work of the organization balances with payroll accounts, employment plans, financial plans, and technical plans. When compared with figures on work accomplished, it makes it possible to calculate the man-hours needed to achieve a given volume of production of a given quality, to work out rough indexes of the time needed for each unit of production, and to measure these against performance on comparable districts. It reveals how much time goes into travel as contrasted with operations, function by function. It discloses whether excessive amounts of time are devoted to paperwork in the office vis-à-vis work in the woods. It permits specialists from higher headquarters to assess the performance of individual functions—that is, to find out not only whether timber management is given its due, for example, but whether each aspect of it gets sufficient attention; not only whether fire control has been accorded adequate time, but whether enough attention has been paid to prevention and presuppression as well as detection and patrol. As an instrument of administrative detection and diagnosis, the diaries are even more sensitive than reports, for they cover field work more minutely. If field officers diverge from the preformed decisions of the Forest Service and the divergence somehow escapes notice by those who review regular reports, it is not likely to be missed when the diaries are analyzed.

Forest Service officials do not designate disclosure of deviation as the chief function of the diaries. Rather, they contend the information in the diaries is needed to enable the leaders to formulate and adjust policies and objectives to what the records show is practicable on the ground; it puts the leaders in touch with reality, and tells them as much about the shortcomings of their own programs and goals as it does about the men in the field. The diaries, they say, are designed for the guidance of the top echelons rather than to force field officers to testify against themselves. And there is no denying the diaries are used for this purpose.

But the fact remains that the diaries *also* expose deviations from decisions issued at higher levels to regulate the behavior of men in the field. The practice of keeping diaries may have been instituted for other reasons, and it may be employed in other connections. Nevertheless, exposure of deviation is one of the consequences, and few members of the organization are unaware of this.

The diaries are kept accurately for the same reasons that reports are not "doctored." In the first place, falsifying them is far too difficult. If inconsistencies between the diaries and work reports were not quickly discovered, then contradictions between what the diaries recorded and what was actually accomplished in the woods *would* soon come to light. Furthermore, discrepancies between the entries in the diary of one forest officer and those of his colleagues, subordinates, and superiors could not be long concealed. In any event, the entries are made throughout the year, so manipulating them for purposes of hiding the truth would take elaborate, long-range planning, and great investments of effort. There is an additional risk created by the use of the diaries as evidence in court in trespass and land-exchange cases; a false report could come to light in that fashion (and weaken the government's case besides). Finally, the costs of accurate reportage of deviation are invariably far smaller than the costs of discovered misrepresentation. All the weight is on the side of accuracy; it is the path of least resistance.

Secondly, there is apparently a feeling of ethical, professional, and organizational obligation to keep the records straight. Members of the Service speak with obvious repugnance of tampering; it is regarded as petty and contemptible conduct, contrary to the traditions and the welfare of the agency. To be sure, few diaries are actually current, as regulations require; except for what one Ranger called "streaks of religion," during which he enters his activities faithfully at the end of each day, most men rely on their memories, aided by brief notes and consultations with co-workers, to fill out the forms for days—or even weeks—during which more urgent business was given precedence. And there is by no means unanimous enthusiasm for the diaries; some men argue that other reports supply all the information that can be gleaned from them. Still, even one Ranger who objected strongly to keeping one admitted that he is conscientious about its accuracy and completeness even though it has disclosed occasional failings for which he was reprimanded. Unquestionably, they are not precise to the minute, but they do reflect fairly closely what actually happens in the field.

The diaries are collected by higher headquarters and analyzed periodically. Current pages, kept in the field for the use of field administrators, are available to visiting inspectors. And they *are* studied by representatives of higher levels.

Along with diaries of their own activities, officers in the field are also required to maintain equipment-use records that are in effect diaries of their equipment. Each piece of apparatus and each vehicle is covered by a log in which an entry must be made every time it is employed. The ostensible objectives are to furnish cost data, and to provide information from which it can be determined whether the equipment is used enough to justify it; in the language of the Forest Service, equipment must "earn" its purchase. But discrepancies between reported use of equipment and technical and financial plans, travel allowances, travel entries in personnel diaries, and work accomplished in the woods are occasionally discovered by comparing equipment records with other documents and inspection reports. Within limits, property records and other reports must tally. They send up warning flags when what happens in the field diverges from what is enunciated as policy at the center.

District Rangers and their subordinates thus leave behind them in time a wake of paper that is a highly visible chronicle of their operations. If they stray from the designated channels, they do not ordinarily get very far before their divagations are disclosed.

Furnishing Overhead Services

There seems to be a consensus in the Forest Service that the Rangers, as field officers, should be relieved of as much clerical detail as possible. The Rangers find paperwork an irksome interference with the performance of their primary tasks, which are in the woods; their superiors would like to free them from their desks to increase general output and to supervise more closely the work of sub-Ranger personnel. Therefore, every opportunity to relieve them of record-keeping and accounting chores is seized. There can be no doubt that the paperwork burden, which burgeoned as the organization grew rapidly, has been lightend considerably in recent years as a result. It is equally certain, however, that the transfer of parts of this burden to higher headquarters (the most common method of resolving the dilemma) provides the Rangers' superiors with additional indices to the character and timing of Ranger activity.

IBM machines, for example, have been installed in all the regional offices, and virtually all bookkeeping and accounting are performed on them. Data coming in from the field units are punched on cards and processed by regional office technicians, and the printed figures are distributed to relevant officials in the regional offices, to the appropriate forest supervisors, and, by the latter, to the Rangers. A glance at the running totals each month informs everyone what is happening on every district, and as a result most Rangers, at one time or another, have had their attention called pointedly to some trend or imbalance requiring redress. The load on field men is eased by the service, it is true, but the service also improves the detection of deviation from promulgated norms.

Similarly, the offices of the forest supervisors do many things for Rangers that take from the Rangers' shoulders burdens they would otherwise have to bear. For instance, on one national forest, although most of the information for timber sales requiring supervisor signature is gathered by the Rangers and their crews, actual calculations from the raw data are made by the supervisor's staff assistants; in the course of liberating the Rangers from much tedious, routine work, the process furnishes the higher level with clues to the efficiency and accuracy of the work in the field. To take another example, on virtually all Ranger districts, the number of vehicles is not large enough to warrant employment of a full-time mechanic, so one mechanic is engaged to service all the vehicles on a forest, moving from district to district on a regular schedule for ordinary maintenance, and remaining on call for emergencies at all times; unwittingly, his reports on his own activities would highlight discrepancies between actual vehicle use and budgetary allotments for this purpose, and would reveal differences between reported use and real use if there were any. In the same way, heavy equipment for construction (chiefly of roads) is kept in forest motor pools because no individual district can make full-time use of it; the equipment rotates among the Ranger districts. In one sense, this imposes a preformed decision on the Rangers, for they must schedule their other work to be ready to engage in construction when the equipment becomes available. In another sense, it makes it difficult to conceal deviations from budgetary and planning allocations for construction, since the equipment operators can hardly help including in their own work reports what goes on in the field. In some regions, construction and maintenance of roads and other permanent improvements are assigned to forest supervisors rather than Rangers; keeping track of the funds, balancing the books, employing and supervising crews and reporting accomplishments are duties of functional specialists on the supervisors' staffs (although the Rangers participate in all planning, and retain authority over all crews in their respective districts). Rangers are thus relieved of responsibilities, but the scope of their independent decision-making is simultaneously narrowed, and their supervisors are more thoroughly informed about the management of the districts than they would otherwise be.

Some services are provided by agencies outside the Forest Service itself. The General Services Administration stocks many kinds of stores and supplies that must be requisitioned from it. The Civil Service Commission furnishes lists of eligible personnel for vacancies, and approves candidates recruited by the Forest Service when no lists are available. The regional attorneys of the Department of Agriculture furnish legal advice and guidance. United States attorneys represent the interests of the federal government in court. Some buildings are under the jurisdiction of the Public Buildings Administration. Arrangements between the Forest Service and other agencies are made chiefly in regional offices. Consequently, though the life of field men is made somewhat easier by the fact that they can obtain the services of many specialists whose work the Rangers themselves would otherwise have to do, or contract to have done, it also puts their superiors on notice about operations in the districts.

Rangers are thus relieved of many chores. While some express irritation with the procedural requirements imposed on them, most of them are happy that they are freed for field work and agree that the services furnished for them are often less costly and more expertly rendered than they would be if the Rangers had to arrange for them. Certainly, the Rangers do not think of every individual and agency that services them as secret policemen bent on finding evidence of wrongdoing. Just the same, when the services *are* furnished through higher headquarters, the upper echelons learn (and sometimes determine) exactly what happens on the districts. This reduces still further the likelihood that departures from preformed decisions will long continue without detection.

Inspection

In the end, however, regardless of how much of their field behavior is described in what the field men tell about their achievements, and in their inadvertent disclosures when they employ staff services, the only sure way to find out what goes on at the level where the physical work of the organization is done is to visit the field and see. This practice has been highly developed and carefully systematized in the Forest Service. Indeed, it has a long tradition behind it; when the agency was formally established in 1905, Gifford Pinchot, the first Chief, created a special corps of inspectors who were to serve as "the eyes and the ears" of his office. When the forerunners of the present regions were set up in 1908, the special corps was eliminated in favor of the system of inspection that now obtains, and the system was refined and intensified as the business of the Forest Service grew by leaps and bounds.[3] From the very beginning, it was customary for

[3] G. Pinchot, *Breaking New Ground* (New York: Harcourt, Brace and Co., 1947), pp. 279-80.

higher headquarters to take long, scrutinizing looks at the field, and the looks have become more searching with the passage of the years.

TYPES OF INSPECTION

The broadest type of inspection—that is, the type that covers the broadest range of activities—is the General Integrating Inspection. It includes everything; no phase of a unit's operations is omitted. The quality of resource management from an over-all point of view is evaluated on the basis of several criteria: the balance among activities, the actual accomplishments measured against the possibilities (given the money and manpower available) as assessed by the inspectors, the degree of adherence to the policy pronouncements (the preformed decisions) of higher headquarters, and the preparation of the organization to handle emergency situations should they arise. The breadth of coverage is supplemented by examinations in depth of projects selected on a random basis, extending from maintenance of files and records and the posting of manuals to relations with the public; from the condition of permanent improvements like warehouses and work centers to the state of expendable tools; from the administration of individual large timber sales to the handling of free-use permits. The visitors look not only for what has been done, but also for opportunities missed, for complaints by field men and for suggestions from them, for excellent work as well as for deficiencies, for imagination as well as for simple compliance, for aggressive pursuit of funds to develop forest resources as well as for maximal use of funds allotted. In short, a General Integrating Inspection is designed to find out how good a job a line officer is doing when his total responsibilities are considered. Taking the whole gamut of national forest administration tasks as its subject, it reveals whether an organizational unit is administered in accord with policy, and whether everything that *could* be done *is* done.

Functional Inspections are narrower in scope but greater in depth than General Integrating Inspections. They normally concentrate on individual functions—timber management, wildlife management, recreation management, information and education, engineering, etc.—and explore in detail the way they are administered. A General Integrating Inspection takes up the balance among functions; a Function Inspection turns to the balance among tasks within a function. A General Integrating Inspection relies on samples and general impressions; a Functional Inspection rests on minute examination and analysis of figures and methods. General Integrating Inspectors strive for "horizontal" sweep; Functional Inspectors aim at "vertical" comprehensiveness.

Functional Inspections are of two kinds: General and Limited. A General Functional Inspection covers all aspects of a function. A Limited Functional Inspection ordinarily singles out a specific task or group of tasks, or even a single project, for intensive inquiry. For example, timber sales administration may be selected from all the activities comprised in Timber Management for special investigation. Fire prevention may be picked out of Fire Control for inspection. A particular game refuge or recreation area may be chosen for examination. Limited Functional Inspections supplement General Functional Inspections; they do not supplant them. They are normally made more frequently, and conducted whenever an official feels there is a need for them rather than at regular intervals.

Fiscal-Administrative Inspections—essentially, audits of fiscal operations and administrative housekeeping functions —do for office management, including accounting and recordkeeping and reporting, what General Functional Inspections do for field operations. Books, files, and records are examined for maintenance according to standard. Manuals and work plans are reviewed to see if they are up-to-date and accessible. Procedures for handling paper work are studied. Diaries are checked for currency and completeness. Reporting promptness and accuracy are evaluated. Thus, not only is behavior compared with preformed decisions by means of reports and records; in addition, reports and records are themselves inspected to ensure the reliability of the data they provide, and to make sure field officers are familiar with the decisions to which they are expected to conform.

Two additional types of inspection are *ad hoc* in character rather than recurrent. They are substantially hearings on major failures of one kind or another, although they are sometimes employed to see what can be learned from unusual accomplishments as well. Boards of review look into the causes and consequences of large-scale, unexpected reverses, such as huge fires; investigations are inspections of alleged misconduct on the part of forest officers. Unlike the other kinds of inspection, these occur only when something extraordinary happens; they focus on the exceedingly unusual.

Whether going along quietly and routinely or beset by catastrophe, members of the Forest Service can be as certain of inspection as they are of death and taxes.

THE CONDUCT OF INSPECTIONS

Each level in the Forest Service inspects the level immediately below it.[4] While the inspectors are regular line and staff officers, there is a special general inspector under the Assistant Chief for Administration in Washington. A General Integrating Inspection of a region by the Washington office will normally be conducted by a team consisting of an assistant chief and the general inspector with such special assistance from any Washington office division as they may

[4] The Internal Audit Division, however, is an exception, for it, on behalf of the Chief, inspects the Washington office as well as the regions.

request; thus the composition of the team will vary with the workload of the region under inspection. A Functional Inspection of a region is generally conducted by a team from the appropriate functional division of the Washington office. Internal audits are conducted by officers of the internal audit staff. Boards of review and investigations are constituted as needed, and their composition depends on the subject to be investigated.

Similarly, General Integrating Inspections of forest supervisors are usually conducted by two-man teams headed by an assistant regional forester and aided by specialists requested of various regional divisions; and representatives of the appropriate functional divisions of the office of a regional forester constitute the teams making Functional Inspections. Regional auditors, assigned by the heads of the regional divisions responsible for fiscal control, handle audits of forests. Boards of review and investigations are composed of appropriate specialists.

General Integrating Inspections of Ranger districts are performed by forest supervisors' offices on most national forests, although they are commonly joined by their appropriate staff assistants as particular functions come under scrutiny. The staff assistants handle Functional Inspections; even the fiscal inspections are performed ordinarily by the supervisors' administrative assistants. There are no boards of review or investigations below the regional level, although officers from the lower levels may be assigned to such duty on higher-level bodies by regional foresters.

General Integrating Inspections, being the most comprehensive and time-consuming, are undertaken the least frequently. Yet even for these, the *Forest Service Manual* requires that each region be inspected at least once every five years, the forest supervisors at least once every four years, and the Ranger districts every three years. Fiscal Inspections are the most frequent; the *Manual* calls for checks on regional offices at least once every 12 to 18 months, on national forests not less than every 15 months, and on Ranger districts a minimum of once a year. Functional Inspections tend to fall somewhere between these in frequency.

When a region is inspected by Washington, or a national forest by a regional office, the inspectors ordinarily visit randomly chosen Ranger districts, for every inspection involves study of field work. Every Ranger interviewed has had the experience of being visited by people from Washington and region offices as well as being checked by their own respective supervisors. Since the number of inspections of one kind or another by supervisors' offices average three or more a year over and above the reviews by higher levels, every Ranger can count on at least several inspections every year; inspectors and functional specialists thus come through with rather high frequency,[5] and the chances of deflections from preformed decisions going undetected are correspondingly reduced.[6]

The pattern of inspection does not vary a great deal from one part of the Service to another. There are no surprise visits, for the objectives of inspection are not so much to catch personnel in the wrong as to find out what is happening in the woods and to train forest officers in the organizationally approved methods of resource management.[7] Be-

[5] In a letter (February 21, 1955) to the Regional Forester of Region 2, the Chief of the Service reported, "Annually we make about 1,600 inspections from the national forest [forest supervisors'] and local research headquarters, 700-800 from regional headquarters, and 100-150 from Washington.

[6] Inspections are somewhat behind schedule on most of the districts studied, but not so far that the frequency of visits to the field by higher levels is significantly below announced standards.

[7] "I want to stress one point," said the Chief of the Forest Service in the letter to Region 2 (*above*, n. 5), "When I say 'inspection' I do not mean 'investigation.' An 'investigation' implies searching for something that's dishonest or otherwise wrong. You usually don't investigate unless you suspect or know that something is wrong. I am talking about something quite different. . . . We always try to recognize good work as well

sides, unless advance notice is given to the units to be visited, there is a chance that the entire staff will be out working when the inspectors arrive, and the inspectors will lose valuable time (and be thrown off their own schedules) while they wait for the return of those they came to see; since inspectors, even at the forest supervisors' level, have many field organizations to examine, they must plan their itineraries in advance and arrange everything beforehand in order to be sure they complete their rounds. Moreover, within the limits of practicability and the requirements of broad coverage, efforts are made to confine inspections to the periods when the workloads of the inspected offices are lightest so as to cause the minimum of disruption in the productive activities of the organization. In any case, surprise is not particularly important in an agency where slow biological processes set the pace; if the things that are wrong are so trivial that an official can correct them in the weeks between notification of an approaching inspection and the arrival of the teams, they are not of great concern, and if they are more serious, it is unlikely that they can be remedied in so short a time. The men to be inspected are therefore told ahead of time when to expect the visitors, and they then adjust their own work plans to accommodate them.

Upon arrival, the inspectors report to the line officer of the unit to be inspected. In the case of a General Integrating Inspection, the line officer will, as a general rule, assemble some members of his own staff, and accompany the inspecing team. Functional inspectors, including auditors, may be turned over to the line officer's appropriate assistants, who show the visitors whatever they want to see and answer their

as to point out needed improvement. We of course try to catch up anything that may not be right, but the attitude is to see how together we can do a better job."

Similarly, a regional regulation describes the objectives of inspection as 90 per cent training, 5 per cent fact finding, and 5 per cent recording and reporting.

questions; even a Ranger may send out an assistant Ranger to accompany a functional inspector from forest headquarters, although he more commonly accompanies visitors from higher levels himself.

The inspectors then ask to see examples of each of the operations, office and/or field as the case may be, with which they are concerned. The group goes over the selected illustrative materials and sites, asking questions, making suggestions, discussing the items, and making notes as they proceed. They travel around examining physical properties in many cases—and not merely by looking out of car windows. They tramp through the woods, test the facilities, study the permanent improvements, watch activities in progress, and otherwise acquaint themselves as thoroughly as they can with the state of affairs on the government's holdings and the surrounding area. (On the whole, however, they do not interview more than a few selected members of the local populace; they rely on other means, discussed below, to find out what local attitudes toward the Forest Service are.) A Functional Inspection on a Ranger district is not apt to last more than a day or two or three at the most, and even a General Integrating Inspection may not take more than a week or so. But the inspectors, many of whom have been Rangers themselves (or, in the case of administrative assistants, have served earlier in their careers in Ranger stations), know what they are looking for and are thoroughly versed in resource management; in a short time, therefore, they can learn a great deal about how a district is run. Rangers occasionally express minor irritations about individual inspectors, but it is clear that they have profound respect for the acumen and knowledge of the inspectors, for their ability to probe field activities, and for the difficulties of concealing realities from them.

Yet all is not tension and friction by any means; indeed, the more experienced Rangers are quite casual about inspection, for they know that only the most grievous mismanagement is likely to get them into serious trouble. The atmosphere of inspection is not one of a trial or even a competitive examination. In the evenings, when the work is done and the notes written up, the inspectors and the inspected gather socially to discuss personal and organizational affairs—such things as shifts of personnel, promotions, retirements, additions to staff, organization policies and strategies and problems—meeting as professional equals rather than as superiors and subordinates, or inquisitors and defendants. The practice of rotation and transfer of foresters, described in the next chapter, combined with the travels of inspectors, acquaints members of the Service in each region with their co-workers; inspection is a mode of communication and of face-to-face contact that helps bind the agency into a unity. Men in the field, rather than fearing inspection, tend to welcome the opportunities it affords them to keep abreast of developments in the organization, to learn the latest rumors and gossip, and to give their own ideas to their superiors at first hand.

The fact remains, however, that the written reports following every inspection are blunt and hard-hitting; criticisms are not softened, punches are not pulled. The following quotations selected from General Integrating Inspections of several Ranger districts indicate the detail into which inspectors delve and the vigor of their comments:

> It is . . . difficult to tell from diaries and other records who is Fire Boss on individual fires. I wonder if the same difficulty is present among your men actually on fires?

> Your attention is called to the handling of Fire #10 . . . P---- picked up smoke at 11:50, checked it and sent F---- to it at 12:03. At 12:38 F---- had scouted fire and radioed he needed help. P---- sent Wardens B---- and M----. His wife got W----'s sawmill crew. The plow was sent. At 2:00 P.M. he notified your office. J----

was there. You were on annual leave that afternoon although it was a class 4 [high fire danger] day.

P---- records that neither you nor J---- had checked with him since 7:45 A.M. You phoned P---- at 3:10 P.M. (J---- got you at home) but instead of proceeding to the fire or sending J---- you told P---- not to send more men unless F---- asked for them. F---- reported the fire controlled at 5:30 P.M.—68 acres—and your men returned to station at 6:30 P.M. This was a bad fire day and the fire should have had your personal attention. I think you were luckier than you deserved to be and not nearly as energetic and aggressive as you should be.

The fire records are confused by fires in the State Zone [the area adjacent to the National Forest under fire protection by the State forestry agency] to which you sent men and to some of which you went yourself. However, there are no [Form] 929's covering these fires . . . If the fires endangered National Forest land, we should have reports on them; if they did not, what were you doing fighting fires there?

W---- is a young, aggressive type of individual, very capable and full of energy. In fact, this latter trait may cause him some difficulties in the future. He makes decisions quickly and expects the permittees to accept them. In some cases if he could 'slow down' and listen patiently to the permittee's story, no matter how irrelevant to the subject at hand, discuss it with him in a slower manner and not be in a rush to get to other business, it is thought better progress could be made. He should practice the art of 'attentive listening.'

The average mileage lapse between grease jobs for the Ranger's pickup is believed to be too great, especially considering the wet muddy roads involved this summer and also since the lapse was over 7000 miles in seven cases.

Generally, the land exchange program has not been pushed to the extent it should be on the District . . . Another area where some exchange work is needed is in the L---- Canyon country . . . Other areas needing special attention are: . . .

Clean up the H---- resort site and effectively barrier

against public use until it can be developed for camp ground use.

The . . . Ranger Station residence needs some window sash replacement in the coming year. Two of the upstairs windows have warped and weathered to an extent where close fit is no longer obtained and replacement will be necessary. There was no indication of rot to the sills or floors, however.

It is recommended that: Office time by Ranger and Assistant be reduced to Regional standards. Impetus be given to Management Improvement program. R---- Tower be repaired as necessary, at earliest practicable date; . . . towerman recruitment to be given high priority. Five year fire occurrence map be brought up to date and kept current.

From analyses of your diaries and service reports, it seems that your functional 030 time [030 is the account number for Timber Sales Administration], 44% in . . . 1953, is sufficient but excessive. A tabulation I made shows well distributed participation with your men on sales work. In . . . 1954, considerable time was spent supervising marking, which was appropriate, considering what we saw last December, but more, in my opinion than should be so spent. As I have expressed it to you, the Ranger should be concerned with what *stands* are cut; the Forestry Aids can take care of what *trees* are cut (if they are well trained, of course).

H----'s audit report revealed some practices not in accordance with instructions. Controls should be tightened to be sure that scaling of logs is done according to instructions. Scaling determines the amount of money due the government. We must protect the interests of the government and at the same time be fair to the purchaser.

Revision of grazing allotment objective statements is a must job before next grazing season.

New tract plans are needed for the following camp grounds: . . .

These excerpts, chosen haphazardly, suggest that no program is too large, no item too small, to draw the attention of the inspectors. And when the inspectors find something below what they understand the standards to be, they do not hesitate to say so, nor to recommend remedial action.

However, they do not hesitate to point up achievements, either:

> A commendable job . . . was done on preparation of the slash disposal plan prior to fall burning. Objectives of disposal were well stated, responsibility for accomplishing the objectives were clear cut, as well as the time for and the manner of burning. Safety was also provided for.

> Ranger R---- has a good grasp of Forest Service objectives, policies, standards, and goals . . . His work is above average in all respects. He did make a change in organization early in the summer which is considered well worthwhile.

> In addition to the organizational change referred to above, R---- was vitally interested in having his subordinates make decisions. This is good leadership and is reflected in the high morale among the employees on the district.

> The district personnel do an exceptionally good job of sale administration and are well qualified for handling the sales business.

> Last fall and spring plantations were observed and found to conform to standards. A good planting job has been done.

> W---- is very interested in work planning and has done a good job of making this style of planning into a useful tool to help him administer the district. The monthly plans were followed reasonably closely and accomplishments are satisfactory.

> Possibly the greatest range management problem of all is I(nformation) and E(ducation). While the Ranger is now doing a good job in this phase of management, much more work and effort is needed.

> Both Administrators [Ranger and Assistant Ranger] of the . . . District are very safety conscious. Actually W---- is above average on this phase of work . . . All of the . . . safety

program items were initiated by the Ranger with no "pushing" by the supervisors' office personnel. A safety-conscious workman and a good safety record are the results.

Indeed, inspectors apparently are under some pressure to keep the reports critical. One Ranger once was a staff assistant to a forest supervisor, and, in that capacity, inspected one function on the Ranger districts of the forest. His report on one district had nothing critical in it—no findings of deficiencies, no suggestions for improvement—because he thought the Ranger had truly done a magnificent job that he could see no way to improve upon. Later on, an inspector from the regional office, seeing this report, made some acid remarks indicating he believed the staff assistant had not performed his inspectorial duties properly; there were veiled hints at incompetence and possibly even collusion. Since the *Forest Service Manual* instructs inspectors not to "waste time on details already being accomplished to a satisfactory standard," and to "save time and energy for more thorough consideration of significant faulty conditions and practices," this is probably not an isolated reaction. Moreover, although inspectors are admonished to "be alert to outstanding accomplishments" and to "recognize these (and) comment favorably upon them," the thrust of the situation tends to drive men toward criticism: Unearthing shortcomings is a way in which inspectors demonstrate their own mastery of their technical fields, their diligence, and their imagination, and it also insures them against attack should a weakness neglected or overlooked be discovered later, or, even worse, result in a serious problem. So inspection reports are frank and unvarnished; they pound their points home.

Inspectors summarize their principal findings for the officers they investigate, and the inspected officers are thus both forewarned of what is to be reported and given a chance to answer the criticisms and thus possibly to have some of them explained or eliminated.[8] If a report nevertheless contains material to which they object, they may submit a written protest; several of the Rangers interviewed have done so on occasion. One Ranger expressed mild annoyance at the appearance in reports of findings not actually discussed with him in advance, and another was somewhat irritated by intimations that he was unaware of deficiencies that any good forester would recognize—deficiencies, in some instances, that he himself called to the attention of the visitors. On the whole, however, the Rangers indicate they believe inspection reports are accurate and fair despite their sometimes painful candor.

INSPECTIONS AND PREFORMED DECISIONS

Inspections thus supplement other procedures for discovering what is actually done in the management of the national forests. What is more, they reveal not only the volume and character and cost of accomplishments, but the quality of performance as well. They divulge how well the forests are being administered. They constitute still another reason the men in the field cannot wander far from the beaten path without detection.

But inspections are not only a mode of detection. They are also a method of communicating preformed decisions to the men in the field, of reducing the ambiguities of previously issued policy statements, and of finding out whether such

[8] In the case of a General Integrating Inspection of a forest, it is common for the forest supervisor and members of his staff to meet with the inspectors and with the regional forester and members of his staff to discuss the report and plan remedial action where necessary; only afterward is the inspection made "official" by a formal letter of transmittal from the regional forester to the forest supervisor. On some forests, parallel procedures are employed in inspections of Ranger districts, but these procedures are by no means universal at this level.

policy statements require revision in the light of field experience. That is, they are an additional technique of preforming Ranger decisions, and they help determine the contents of such decisions.

For inspectors do not merely note violations of policy pronouncements and suggest in general terms that the field men look up the appropriate provisions and figure out how to conform. Rather, they indicate quite precisely what is to be done—what neglected projects should be undertaken, what activities should be reduced or halted or expanded or intensified, what procedures should be improved or corrected.[9] They direct and prohibit action. They interpret authorizations and plans and budgets. They clarify ambiguous statements. In so doing, they claim merely to explain what policy statements and rules and regulations mean; this, in fact, is why they are said to be engaged primarily in training. Yet it is clear that they fill in whatever interstices may remain in the fabric of preformed decisions; they tighten the weave.

Not every such elaboration of the body of administrative issuances is initiated by the inspectors. To be sure, in the written documents, in the conversations during tours of the physical facilities, in the discussions of tentative findings and recommendations, and even in the informal social evenings, the visitors volunteer their ideas on a great many matters even in the course of a couple of days. But some of their advice and suggestions are requested by field men unable to interpret an instruction, or uncertain about how to resolve apparent contradictions between various provisions, or anxious to find impressive support for an interpretation on which they have been overruled by someone else. The inspectors do not simply impose themselves on the field officers; the field officers take whatever advantage they can of the presence

[9] See note 7, *above*, for example.

of representatives of higher levels by inviting interpretations and elaborations of the rules. In a sense, this is a form of clearance. In any event, the Rangers elicit by their queries some of what is told them by their superiors.

At the same time that inspections increase the volume and specificity of decisions flowing *to* the field, they afford the Rangers opportunities to influence the formation of some of the decisions they will be expected to abide by. For the Rangers take advantage of the personal contacts with the inspectors to voice their complaints, their needs, their preferences, and their aspirations. If objectives are unattainable with the funds available, they point this out. If a prescribed procedure is excessively burdensome, they let the inspectors know. If they see potentialities in the management of their districts that call for amendments or additions to existing orders, they do not hesitate to urge them. Admittedly, not all of these individual criticisms and suggestions produce results, but some—especially when they originate more or less independently on a number of field units—do find their way into the flow of changes in, and additions to, the regulations of the Service and the Department of Agriculture, and, on rare occasions, into Presidential orders and into statutes. In any case, the performance and reactions of the men in the forests guide the agency leadership, helping them ground their decisions in what is physically possible. There is a flow of communications upwards as well as downwards within the inspection process, and it may be presumed to guide and limit to a small degree the contents of the rules and orders Rangers are called upon to observe.

Yet while inspections generate elaborations of authorizations, directions, prohibitions, and other preformed decisions, and contribute to the substance of those decisions, their *distinctive* function is to uncover deviation by field men from the behavior prescribed by the organization. Even if there were no inspectors, orders would flow out to the field, and reports and reactions would flow back to the center. But leaders would then be dependent entirely on evaluations of field accomplishments by the very men who did the physical work, men with heavy stakes in making their performance look as good as possible. Inspectors from higher levels, checking the field work, furnish more disinterested judgments as to whether or not the work conforms to policy pronouncements. In the last analysis, this is the rationale of inspection.

Hearing Appeals by the Public

Every Ranger interviewed has at one time or another received an inquiry from higher headquarters, or been visited by a representative of a higher level, as a result of a complaint by some private person denied a claim of one kind or another. The complainant may approach a forest supervisor, a regional forester, the Chief himself, or a United States Representative or Senator. The issue may be land exchange, the terms of a special use permit, timber sales policy, prosecutions or suits for trespass, enforcement of the terms of agreements or contracts, or any of the other functions of national forest administration. The right of appeal is guaranteed by custom, the Constitution, legislation, and regulations; citizens are not hesitant about invoking it. So no Ranger can ever be sure that the people with whom he deals in any connection will not appeal over his head when they are displeased with what he does, not even when the matter in question seems quite minor.[10]

[10] Rangers dislike appeals cases, and do not encourage them, because they are time-consuming and unpredictable. Nevertheless, the Rangers often give appellants information about their rights, proper procedures, and remedies that makes the filing of the appeals easier. (One forest officer suggested that even offers of help in the preparation of appeals is sometimes a subtle mode of discouragement. "It is not uncommon," he wrote, "for people to back down, having used the threat of appeal as a negotiating maneuver. A sincere offer of help with an appeal effectively

Most cases arise not because Rangers *fail* to adhere to the preformed decisions of the Forest Service, but because they *do* conform. Their actions are consistent with announced policies, and appellants objecting to individual acts or decisions of Rangers are often really opposing a leadership policy. Consequently, most appeals are settled in favor of the Rangers.

However, this is by no means automatic. A Ranger can be sure his superiors will not back him up out of organizational loyalty if he is on unsound ground; if the act that provoked the complainant to appeal is not in accord with preformed decisions of the agency, the Ranger can expect a reversal (which suggests a tacit negative appraisal of his judgment) or, if the fault is serious, a reprimand. The consequences of an appeal are never certain. The only thing that is sure is that no complaint (other than obvious crank and "crackpot" letters) will be ignored—particularly if it is routed through a Congressman.

Because of the uncertainty about what might become of an appeal case and where the case might lead, and because an officer's best justification for an action that generates an appeal is that the action conforms to rules and regulations, field officers have strong incentives to stick as close as they can to the preformed decisions issued by the Forest Service—particularly if they suspect that a person with whom they are conducting negotiations is likely to object if his demands are not completely fulfilled. Under these conditions, hostility breeds fastidiousness; the Rangers grow exceedingly careful

kills the threat as a maneuver." But of course this is not always the situation.)

Frequently, the existence of a policy directive gives the Rangers a refuge from local pressures; if they cannot persuade a complainant that an action is justified, they can—and do—claim they have no discretion in the matter. This protects them from local pressures; the appeals procedures are thus welcome devices for deflecting demands to higher levels. Despite the uncertainties associated with them, they are occasionally blessings to Rangers as well as burdens.

when storms are brewing. Thus, appeals procedures tend to discourage deviation by field men from policy pronouncements.[11] And they do this by bringing to light deviations that might otherwise go undetected. They differ from the other methods of detecting deviation because they are intermittent and haphazard in their operation; whereas the other methods depend on their thoroughness and comprehensiveness, the effects of appeals grow out of their uncertainty. They are unpredictable in timing, substance, and origin; they offer no systematic coverage of periods, functions, or areas. In addition, they develop more frequently around instances of observation of standards by Rangers than around violations of standards. Still, they compel higher administrative levels to check what goes on in the field, flagging for special attention events that might otherwise have gone unnoticed. Furthermore, field officers can never be sure what will erupt as a major case. The net result is a general disinclination to depart from established policy.

Movement of Personnel

Added restraints on deviation are imposed by Forest Service transfer and promotion policies. Professional per-

[11] One expert reviewer of this manuscript, however, contends that the leaders of the Forest Service become extremely sensitive to controversial appeals during periods when the agency is attacked or threatened, and that they tend to shrink from rigorous enforcement of the rules when such action might arouse powerful opposition or alienate strong allies at critical moments. But the *field* men rarely make such decisions on their own; adherence to the rules is their chief self-protection. Even in the limited sample covered by this research, there were two reported instances in which field officers refused to sign papers embodying decisions they considered to be in conflict with the rules, and insisted that their superiors, who made the decisions, actually sign them. In other words, appeals procedures and other detection practices do discourage deviant *field* behavior—sometimes more effectively than some officials apparently would prefer.

sonnel are transferred from station to station rather frequently, particularly in the early stages of their careers; and promotion, too, generally means a change in locale as well as in administrative level. This policy, described more fully in the next chapter, was not instituted for the purpose of bringing to light failures to adhere to preformed decisions. Nevertheless, it has this effect.

For when a man is reassigned or promoted, he is replaced. If, in spite of all the methods of detecting deviation, he has succeeded in managing his unit in a fashion inconsistent with Forest Service programs and concealing the inconsistencies from his superiors, he certainly will not be able to hide them from his successor. While there is a fraternal spirit among forest officers, and they are not eager to make trouble for their colleagues, neither are they willing to take the risk of suffering for the deficiencies of their predecessors, or of becoming accessories after the fact. In any case, as soon as they request funds to remedy the faults, or alter the emphases in the administration of their units, the facts are brought to the attention of their superiors. The behavior of the violators is inevitably divulged.

Every man has an opportunity to decline a proposed transfer or promotion. But no man who has been in the Service for any length of time has not served on several administrative units at the very least. Eventually, unless they have accumulated a great deal of seniority and no longer care to advance, forest officers will be moved, sometimes when they least expect it. Therefore, they must keep their affairs in order all the time, for they ordinarily cannot correct the deviations of several years in a moment, nor can they secrete them from those who follow. Steady compliance with announced policy is the easiest way to avoid difficulty. As a means of inducing men to conform and of exposing noncompliance, movement of personnel exerts a constant integrative pressure.

Sanctions

The likelihood that deviation will be discovered by one or another of the means of detection is normally enough to discourage it—not because such discovery is commonly followed by punitive action, but because it reflects adversely on the officer responsible for the deviation. At least a part of the penalty seems to be internal—a combination of guilt feelings, embarrassment, and anxiety about a possible suggestion of incompetence. Many of the behavioral restraints on the individual seem to be of this character.[12] Hence, detection alone is generally enough to induce compliance with preformed decisions, which is the main reason the Forest Service is able to take the position that the objectives of the instrumentalities of detection are primarily fact-gathering and training.

Nevertheless, a broad array of formal sanctions is at the disposal of Forest Service leaders for use if the ordinary methods do not succeed. They range from reprimands (oral for minor failings, written and inserted in personnel folders for more serious ones) through punitive demotion, suspension, and discharge. For irregularities of a criminal nature, a forest officer may be indicted and prosecuted in a court of law. In practice, the milder sanctions are the more common by far; suspension and discharge of foresters have been exceedingly rare, and there are no cases on record of all or any becoming involved in criminal activities. The fact that severe penalties are uncommon does not mean they are regarded by members of the Forest Service as dead letters; on the contrary, in the course of conversation, a number of persons voluntarily speculated on how far they could depart from a particular norm before they would be fired. The impact of formal sanctions on field behavior, in short, is

[12] In this connection, see the section on "Building Identification with the Forest Service" in Chapter VI.

greater than might be inferred from the infrequency with which the major ones are employed. Clearly, they reinforce the internal restraints on the behavior of field men.

In addition, there are a number of agency practices having disciplinary consequences, although they are not regarded primarily as modes of discipline. These are personnel management techniques, described at length in the next chapter. Designed to identify capable individuals and advance them in the organization in order to strengthen the agency, they have developed into a system of reward and punishment closely related to the degree of adherence to preformed decisions. Performance ratings, placement, and promotion are the principal devices; whether a Ranger is rated as excellent or mediocre or unsatisfactory on the merit rating forms, whether he gets an assignment bearing authority and responsibility or a narrowly circumscribed position, whether he is passed over for promotion or moved slowly ahead or advanced rapidly is determined in large measure by what the regional forester and assistant regional foresters think of him. Adherence to agency standards is one of the main criteria of judgment. Judgments usually are based not on any single act, but on an individual's whole record. Ultimately, then, the men whose administrative behavior fits the desired pattern are the ones who win the rewards, while those who do not conform feel the weight of official disapproval.

In sum, detection *per se* is a corrective, both because high-level officials can then tell field men what is wanted of them and because it is in a sense a penalty in itself. It is supplemented by the rewards and penalties of personnel management, which are distributed more or less continuously. And the whole system is supported by a familiar assortment of formal sanctions, to be applied with appropriate severity for more serious or repeated digressions. Deviation cannot long remain hidden under these circumstances, and, once exposed, cannot long continue.

Feedback and Correction

To resort to an analogy currently in fashion, the Forest Service may be described as a system that has been "programmed" and equipped with a built-in "feedback loop." The program is the set of instructions, the series of sequential steps embodied in the preformed decisions, prescribing the behavior of the members of the organization. The feedback mechanism consists of all the modes of scanning actual behavior in the field and detecting departures from the program; this information flows back to the leaders of the agency, who, if the very fact of detection does not result in corrective action by the personnel involved, then apply sanctions to induce them to conform. (If lesser sanctions fail, the individuals responsible for the deviations are simply replaced in much the same fashion as a defective part in a mechanical or electrical system.)

Such analogies can be misleading because the human range of response to programming is infinitely broader and more complex than the responses of mechanical or electrical contrivances. But they serve to illuminate the great problem of organization leaders: the lack of uniformity and predictability in the elements of the system whose operations the leaders seek to control. Whereas an engineer can assume the parts he handles will respond always in the same way to the same signals under given conditions, and that two parts of the same kind will function in the same way, the administrator is confronted with people who vary over time—their moods change, their likes and dislikes fluctuate, their enthusiasms wax and wane, their attention wanders, etc.—and who differ from one another in the way they react to identical situations. Achieving integration in machines of the most complicated kind is simple compared to achieving it in associations of human beings.

This explains why influence exerted by administrators upon the members of their organizations is not enough to integrate their activities; no matter how detailed and specific the programming, how elaborate and sensitive the feedback mechanism, how forceful and relentless the techniques of correcting deviation, the members of human associations still tend to respond to other influences as well and therefore to behave in ways the leaders do not expect or want. Under these conditions, it was inevitable that managers should try to control what goes on *inside* each individual organization member, to get them to do *of their own volition* what the managers want them to do, and to equip them with the resources of skill and knowledge for those duties. Complete success in such an enterprise would reduce to negligible proportions many of the uncertainties in the system, particularly in comparison with the relatively crude modes of external influences. Even partial success might at least make personnel more receptive to instructions from agency heads, and less amenable to forces that would lead them from the managerially prescribed path. The methods of manipulating the internal characteristics of Forest Service personnel are described in the next chapter.

CHAPTER VI

DEVELOPING THE WILL AND CAPACITY TO CONFORM

Selecting Men Who Fit

If a man is willing, or even eager, to carry out the preformed decisions of his superiors, but lacks the technical knowledge and the practical skills to do so, the decisions may never be executed at all, and certainly not properly. If he has the knowledge and training, on the other hand, but is vehemently opposed to the decisions he is called upon to execute, the results may be equally disastrous from the point of view of his leaders. It takes both the will and the capacity to conform for a member of an agency to do his job as the leaders of the agency want it done. The Forest Service therefore tries to get people who have both.

The campaign begins with recruiting efforts directed at young men in the last year of high school or the early years of college; pamphlets, films, and speakers supplied by the Forest Service call their attention to the opportunities and attractions of forestry as a career and advise them how to get further information on training and forestry career possibilities. The campaign does not paint an unmixed picture for the potential recruits, however; while it points up the satisfactions of work in forestry, it also indicates the hardships as well. In this respect, it follows the lead of Gifford Pinchot, who, many years ago, adopted a policy of telling the whole story in a plain, unvarnished fashion:

> Young men began to ask about Forestry as a career. Most of these youngsters I discouraged on the ground that if a boy had the stuff in him to make a good forester he would keep at it anyhow. I told him Forestry means hardship and hard work, much responsibility and small pay, which was the cold fact.[1]

Almost half a century later, Pinchot was still pursuing the same policy. Addressing American soldiers in England after World War II, when they were weighing occupational choices before them as they contemplated their return to civilian life, he told them:

> I do not urge any man to take up forestry as a profession. Unless his own love of the woods, his eagerness to work in the woods, drives him to forestry, he had better spend his life in some other way.
>
> For the man who is by nature fitted for forestry, no other profession offers, in my opinion, such promise of usefulness and such opportunity for a happy life. To those of you who are sure beyond question that forestry is the profession for you, I offer my sincere congratulations and my very best wishes.
>
> The national need for the forest and the opening for forestry and foresters was never as great in America as it is today. The risks of the profession are less than they were when forestry in America was new. But the destruction of the forests since those early days, and especially during the war and the aftermath of the war, give to forestry an indispensable usefulness which justifies any man who loves the woods in taking it up.
>
> I end as I began: Unless you are sure, let forestry alone.[2]

[1] G. Pinchot, *Breaking New Ground* (New York: Harcourt, Brace and Co., 1947), p. 64.

[2] "From Gifford Pinchot to Students of Forestry at Shrivenham American University," *Journal of Forestry*, Vol. 45, No. 5 (May 1947), p. 353.

Weakly motivated men were thus advised to turn to other pursuits. To a considerable extent, those who persisted were self-selected, a rather dedicated group prepared to accept whatever the profession had to offer.

Today, the same practice, a bit modified and softened, but still strikingly candid, obtains. The Service still offers almost as much discouragement as encouragement. For example, potential candidates are told:

> Many persons still have only a vague idea of the kind of life the forester really leads. Young men are often attracted to the profession because of the prospect of outdoor work. They are fond of camping in the open and of hunting and fishing.
>
> One who is considering such a career should remember that the forester in his fieldwork sometimes must endure hardships that sportsmen do not encounter. Spending considerable time in the woods as part of one's regular business is quite different from camping out for a few weeks on a vacation.
>
> . . . If he shows outstanding ability, the young forester may find the apprenticeship period a short one, although as the number of foresters increases and competition becomes more intense, the training period may be expected to lengthen.
>
> The young forester is apt to have his headquarters shifted frequently, somewhat like the civil engineer. The places to which he is assigned may not always be the most desirable from the standpoint of personal comfort or convenience, especially for family life. Because of this shifting about, he may be unable for some time to establish a home. On the other hand, if he is an able man, he may ultimately advance to a position that will give him more permanent headquarters and greater opportunity for home life. He must not count himself secure, however, against a change of working field which will necessitate removal to a new place. . . .
>
> Even in the higher positions, whether in Government or private work, the forester may have to spend a good deal of time supervising or inspecting actual field operations. Trips away from his headquarters may be only for a day or so, or they may be for several weeks. In some positions such travel often includes long, hard journeys by horse and pack train.

> Frequently, it means rough walking, and sometimes days of slow and laborious progress by snowshoe or canoe. Even with the greatest possible extension of good roads, much of the forester's travel for many years to come will necessarily be arduous.[3]

Similarly, in talks before the Society of American Foresters, the professional association of foresters, Forest Service officers have presented the facts with stark candor:

> The Forest Service organization presents the form of the traditional pyramid, with necessarily decreasing opportunities for advancement toward the peak. The relatively low rate of turn over, perhaps emphasizes this. . . . There is always a demand for outstanding men to fill responsible positions. Nevertheless, for many years, it has appeared inevitable that the Ranger's job would be the highest position to which many technical foresters entering the Forest Service can advance.[4]

Who, after these warnings, chooses forestry in general as a career, and sets his sights on the Forest Service in particular as an employer? Logically, it would seem to be those who highly value the work itself, and to whom the agency as an organization is attractive. Recruiting publicity tends to deter the impatiently ambitious, the seekers after the easy job and the comfortable and stable life, and the men who grow restless at the thought of positions within the framework of a large organization, with all the administrative burdens and frustrations this entails. To be sure, it is by no means clear that Forest Service publicity is a major factor in the career choices of many foresters, including those in the Service, but it seems safe to assume that at least some of those who enter schools of forestry were exposed to such publicity and were not disheartened by it and doubtless already would have a predisposition to accommodate them-

[3] *Careers in Forestry*, Forest Service, Miscellaneous Publication No. 249, U. S. Department of Agriculture (Revised August, 1955), pp. 3-4.

[4] *Proceedings of the Society of American Foresters, 1947* (Washington, D. C.: Society of American Foresters, 1948), p. 45.

selves to the demands made upon members of the Forest Service. Indeed, *all* the men choosing professional forestry as an occupation may be said to have demonstrated by that choice a set of interests and attitudes indicative of a degree of receptivity to the requirements of life in the Forest Service. Some among them, however, will doubtless have made their career selections knowing *specifically* from Forest Service literature what they can expect of the agency and what it will expect of them.

Willingness to conform is in this sense employed as an initial criterion of selection, a standard applied to themselves by the young men crossing the threshold to professional training. All of the Rangers interviewed declare they entered forestry with no illusions about the work. They started with yearnings to work in the field, whatever the hardships, and with profound respect and admiration for the Forest Service and its officers. A couple turned down excellent opportunities elsewhere in order to be associated with the organization and its program. Several had had personal experience through relatives or temporary work that acquainted them beforehand with the nature of the tasks they would be required to perform. In short, they had many of the qualities and attitudes the Service wants in its men, and they were ready even prior to their professional schooling for the demands that would be made upon them later on.

The formal schooling of foresters in the United States concentrates heavily on the technical aspects of the profession —on biology, ecology, silviculture, and forest economics. There are but thirty-eight colleges and universities in the country that offer instruction in forestry at the professional level (twenty-seven of which are accredited by the Society of American Foresters). It is estimated that 90 per cent of the degrees awarded (chiefly the Bachelor of Science in Forestry, but including a substantial number of Masters and a sprinkling of Doctorates) are in general forestry, while the remaining 10 per cent are in special fields like wood technology, range management, wildlife management, forest recreation, and general conservation.[5]

Since so much of the work of a district Ranger is administrative, involving the reconciliation of technical and managerial problems, no one contends a graduate of a forestry school, even if he holds an advanced degree, could step fresh from his college into a Ranger position. On the other hand, foresters have a common set of technical tools and techniques, a common lore and body of knowledge, so the Forest Service can take for granted many things about the way they would handle property under their jurisdiction. Over 90 per cent of the more than 4,000 professional employees of the Forest Service are foresters; the existence of a widespread consensus on technical matters within the agency is therefore not surprising. In other words, many decisions and actions taken in the field are implanted in these men during their preservice education; appropriate behaviors are in this sense "built into" them. Their receptivity to agency directives is thus produced not only by the constraints upon them, but also by the education and training which results in their wanting to do of their own volition what they are formally required to do. The Rangers interviewed all hold Bachelor's degrees in forestry, and one has a Master's as well, although several of them entered the Service a quarter of a century ago, when college degrees were less usual than they are today. As is true of most forestry students, they all had some field experience, some of it with the Forest Service, gained for the most part during the summer intervals between school years, before they acquired their degrees.

[5] H. Clepper, "Forestry Education in America," *Journal of Forestry*, Vol. 54, No. 7 (July, 1956), pp. 455-57; G. D. Marckworth, "Statistics from Schools of Forestry for 1957: Degrees Granted and Enrollments," *Journal of Forestry*, Vol. 56, No. 2 (February, 1958), pp. 129-35. See also, "Colleges and Universities in the United States Offering Instruction in Forestry (November 1, 1957)," compiled by the Society of American Foresters.

The Forest Service takes pains to induce as many members of this professional reservoir as possible to apply for employment in this agency. All professional positions in the agency are in the classified civil service; recruits enter at the lowest professional ranks by passing the Federal Service Entrance Examination (Forester Option), a written competitive test.[6] Forest Service personnel officers in Washington and in the regions maintain contact with educational institutions, notifying them of approaching examination dates, reporting on the results of examinations, helping in studies (for example, of academic standing and performance on the tests), and appearing before student bodies both as technical lecturers and recruiters. This provides them with a steady flow of candidates. Recruiting is not confined to schools of forestry, for the examinations are open to all who have "subject credits . . . equivalent to those required for a Bachelor of Science degree in forestry," but the most intensive efforts are directed at them, and the overwhelming number of candidates and appointees to professional vacancies come from this source.

The percentage of applicants who pass the entrance tests generally depends on the number of vacancies to be filled. Variations in the percentage of applicants who pass the examinations result not from variations in the difficulty of the tests from year to year, but from changes in the passing grade. When the number of vacancies to be filled has been determined, the passing grade is set at a figure that admits about twice the required number of applicants (to allow for declinations). Thus, the ratio of successful to total applicants varies a good deal.[7] In a year of many applicants and few vacancies, the percentage of successful candidates is low; in

[6] A few professional foresters come in through examinations for subprofessional positions (Forestry Aid and Fire Control Aid) and become eligible for promotion to professional grades when they complete their period of probation.

[7] This is a standard practice in federal entrance-examination scoring.

a year of few applicants and many vacancies, the percentage soars. (Of course, there is never a time when everybody gets through, regardless of the state of the market; there are absolute as well as relative standards.) Before America's entry into the Second World War, when there were more applicants than jobs, the figure was quite low (18 per cent in 1940, for example), but it climbed rapidly (to an estimated 85 per cent) when the postwar industrial demand for foresters far exceeded the supply; it dropped again (to about 40 per cent in 1950) when the demand was temporarily satisfied, and has risen again in recent years (partly because increasing opportunities in private employ have drawn off many potential candidates).[8]

Consequently, it is no longer true, as it was in the past, that the Forest Service gets only the pick of the crop in terms of academic record, particularly since the examination for foresters, like all other options of the Federal Service Entrance Examination, has been changed from a subject-matter test to a test of general mental ability. That is not to say anything is wrong with the examination; this question is irrelevant to the discussion. It does mean, however, that

[8] "Traditionally the Forest Service has been able to hire the cream of the graduating class of foresters each year through most of the first half of this century. Graduate foresters competed for federal jobs. Today the Forest Service no longer enjoys an employer's market, indeed this situation has now changed to the point that the Service gets about 20% of the forestry school graduates. Private industry, the state forestry departments and other federal agencies are all in the market for foresters. The result has been a deficiency in foresters for the Forest Service. Beginning in 1954, the Service was unable to fill all vacancies. . . ." W. A. Elkins, *Professional Manpower for the Forest Service* (mimeographed by the Forest Service, 1957). The acute shortage of foresters for the Forest Service was somewhat relieved by the recession of 1958, during which private industry absorbed a smaller proportion of the graduating classes of the schools of forestry than it had for several previous years. But the basic problem—the willingness of industry to pay much better salaries for the ablest young foresters—continues to weaken the competitive recruiting position of the Service in terms of both quality and quantity.

examining practices today place heavier reliance on professional training vis-à-vis test scores for evidence that the men admitted to the Service have the desired technical premises of action "built into" them. Self-selection by career choice, coupled with schooling, do more to assure employee receptivity to agency direction than does testing in a full employment economy.

However, once men are taken into the Forest Service, the members of the agency, who have relied up to this stage on others (the applicants themselves, and their teachers) to provide some of the sought-after propensities in candidates for permanent appointment, take a hand in the process of selection. Under civil service regulations, entrants into the Federal Service must serve a year as probationers before they receive "status"—i.e., procedural protections with respect to discharge or demotion as long as the jobs they occupy remain in existence, and certain preferences with respect to re-employment if their jobs are abolished; during that year, the hiring agency may discharge them at will. Appointing officers (regional foresters for the lower grades of the Forest Service) are thus able to judge the qualifications, attitudes, and personalities of newcomers on the job, and to filter out those who, although they chose forestry as a career and came through the training and the examinations, seem for one reason or another to lack the willingness or the capacity (or both) to conform to the preformed decisions of the bureau leadership. In point of fact, relatively few men are dropped on agency initiative during probation, chiefly because the earlier phases of the selection process turn back most of the men who would not fit. But as many as 10 to 15 per cent withdraw from the organization (some, it is true, at the suggestion of Forest Service line and personnel officers; most, on the other hand, on their own initiative) within the first three years after initial employment. Those who make the grade and stick to the Forest Service are therefore men who know the agency and are not at odds with its goals and methods. They have shown an intrinsic readiness and ability to conform.

Post-entry Training

From this point on, the Forest Service undertakes deliberately to intensify both the readiness and the ability: To what the men who survive the screening of the selection process bring with them—have "in" them, one might say—the officers of the agency attempt to add by training.

Some of the training is conducted on a group basis. As young junior foresters—men appointed after passing the entrance examination—come into the Service, they are assembled for orientation courses in their respective regional offices to get an introduction to the history and the mission of the Forest Service, its place in the federal government, the essentials of federal employment, and the nature of the ground-level job and its importance in the execution of Service-wide policies. For several days, the junior foresters stay together, listening to talks by specialists on the staffs of the regional foresters and by selected speakers from outside the Forest Service, and attending discussion and question periods.

In addition, conferences, schools, and training camps are conducted as refresher courses, or to introduce new policies and procedures, or to provide instruction in the handling of special problems and the application of new scientific discoveries for which pre-service training and past experience are not adequate. Some of these are Service-wide, some are regional, but, while Rangers appear at a few of the latter, the ones to which they are most frequently exposed are those at the forest supervisors' level. The training histories of the Rangers studied are by no means identical, but the array of more or less formal courses of practice and instruction which they, considered as a collectivity, have taken is imposing, including as it does general conferences or encampments for review of the whole of the Ranger's job; functional conferences and field studies for wildlife management, timber management, and other phases of Ranger district administration; schools in administrative management; and courses in job instruction, job methods, and job relations for supervisors, developed during World War II and adapted for use in the Forest Service. And every Ranger attended at least three such formal programs of instruction during his early years in the Service.

But schools and conferences, in spite of the emphasis placed on them in the Forest Service, constitute only a small fraction of the training conducted in the agency. For according to the *Manual:*

> It is the responsibility of each supervising officer to consider annually the training needs of his or her immediate subordinates and to initiate action to provide the opportunity for as much of the needed training as is officially justifiable or practicable.

Allowances for formulation of training plans for each professional employee are included in the financial and workload figures for every level. Training bulletins, film strips, motion pictures, and other materials are provided by personnel officers in Washington and the regions. Training is checked in the course of inspections. Every supervisory officer is continually urged, consciously and deliberately, to train his subordinates. On-the-job training is unquestionably the largest single element in the Service training armory.

Except for new men, most on-the-job training is informal in practice. Extensive plans are drawn up for probationers moving into Ranger districts at sub-Ranger levels. They are instructed at some length by the Rangers, given exercises and formal reading assignments, and conducted on tours of their respective districts. Very soon, however, they are sent into the field on their own, and their work is later checked by the Rangers, who point out specifically what is wrong and why. Assistant Rangers are often asked to draft letters for the signature of the Rangers, who then discuss with them the problems of strategy and authority invariably involved in such written contacts with the public and with higher echelons, explaining each change and the reasons for it. Assistant Rangers are often invited to attend discussions between the Rangers and their clienteles or superiors. In the three to ten years it generally takes a probational forester to reach command of his own district, he is subject—by careful planning, not just by accident—to concentrated training in the form of close supervision. The closeness of supervision is not relaxed, and the breadth of his discretion and the burden of his responsibility are not increased, until he has demonstrated that he knows what his superiors want and that he is capable of doing it.

When a man becomes a Ranger, the intensity of his on-the-job training slacks off a great deal. Still, such training does not cease altogether. Every contact, every request for advice or clearance, and every inspection is an opportunity for still more, much of it chance, but some of it intentional, as officers from higher offices discuss orders, actions, and findings with the Rangers. Whatever the purposes of these procedures, the consequences are to clarify for the Rangers the expectations of their leaders and to impart to them the appropriate way to fulfill those expectations. All training is communication, and, while it is not true that all communication is training, it is clear that far more organizational communication has the effect of training than bears the formal designation. The process thus never stops.

Both technical information about silviculture and agency procedure, on the one hand, and more general information about Forest Service objectives, philosophy, and self-justification, on the other, tend inevitably to become intertwined in all training. But the stress is more heavily on the former than the latter after initial indoctrination. Training is more commonly focused on the "how" of the immediate job rather than on the "why" of the Service; it is essentially an effort to inform rather than persuade, to explain rather than arouse enthusiasm for the organization or devotion to its "cause." On occasion, it takes the form of explicating an agency "line," as was the case with a training memorandum from the Chief in 1946 with regard to his proposals for pressing for regulation of cutting on private lands; the memorandum set forth the answers to 55 questions on the subject that members of the Service were presumably to treat as the official Service position.[9] More frequently, the importance of the Forest Service job to the welfare and strength of the nation and its people are pointed up in passing, reaffirming the "faith" of the members of the Service in the work in which they are engaged. On the whole, however, the tone is explanatory, not exhortative. Other modes of influencing behavior may do more to develop the *will* to conform to preformed decisions, although training helps; training builds primarily the *capacity*—the skills, knowledge, and factual premises—that facilitate and encourage adherence.

At the same time, another aspect of training operates as a screening device, selecting for advancement men with the "proper" qualifications *and* motivations. For, in addition to required pre-service and in-service training, they are urged to equip themselves, often on their own time and even at their own expense, to move ahead:

> Forest Service managers and executives are given an opportunity to make a continuous study of public administration and administrative management. They are provided with lists of the better study references. From time to time outstanding

[9] Memorandum from the Chief, "Questions on Forest Regulations," April 1, 1946.

> books and articles are routed to them for study. They are encouraged to enroll in special short-term university courses such as those given in Public Administration at the American, Georgetown, Michigan, Montana, and New York Universities. A few are selected for specialized training in the Intern Management Program sponsored by the Civil Service Commission in Washington, D. C.
>
> To aid men in the more advanced phases of managerial work, special emphasis is placed on group conference participation. Each man is given opportunities to become skilled in conference leadership, in committee work, in panel discussion and in general conference participation . . . As a result of the conference training, a large percentage of the managers and executives become skilled in the art of conference leadership and as a by-product, become accustomed to public speaking.
>
> Special significance is attached to both public speaking and writing. Occasional courses on these subjects are sponsored officially. In addition, men are encouraged to participate actively in speaking clubs and formal adult courses offered by high schools and colleges . . .
>
> The Forest Service has learned that the better executives usually participate actively in civic, fraternal, and religious organizations on their personal time . . . No official stand is taken with respect to such effort but most men do voluntarily take advantage of the opportunities offered for improving their abilities, particularly in the field of public relations.[10]

The education *per se* doubtless does prepare them for the work they will have to do. It does enlarge their capacities to perform their jobs. But those who acquire this training voluntarily, who at some cost to themselves avail themselves of agency-sponsored or -approved programs, indicate by that very fact that they *choose* to fit themselves into the agency pattern in order to advance. It tends to suggest a good deal about their motivations. While, conceivably, men may engage in voluntary training with no aim but to advance them-

[10] C. K. Lyman (Assistant Regional Forester, Region Seven), "Managerial Development in the U.S. Forest Service," unpublished manuscript.

selves, the fact that they do so might be taken as an indication of their tendency to act as the organization wants, to ingratiate themselves with their superiors, regardless of how cynical their attitudes toward the training itself may be. In any event, exposure to the training may well have a more profound effect on them than they realize.

At the very least, then, post-entry training in the Forest Service expands the abilities needed to conform to preformed agency decisions. It also tends to reinforce dedication to the agency and its objectives, although probably less so than some other techniques (discussed below). And, in its voluntary aspects, it helps identify men whose eagerness to advance manifests itself in the will to do what the leaders recommend, let alone direct.[11]

Building Identification with the Forest Service

In addition to picking and advancing men likely to be receptive to communications from the leaders of the agency, and to "training into" these men the capacity and willingness to adhere to preformed decisions announced by the leadership, the Forest Service enjoys—largely as a result of its deliberate efforts, but partly in consequence of fortuitous circumstances—an environment conducive to an almost automatic tendency to conform to those decisions. That environment is a set of conditions promoting identification[12] of the

[11] Of those interviewed, one had taken a Master's degree in forestry and one a course of training in Public Administration at the Littauer School at Harvard. It is probably more than a coincidence that both have moved ahead rapidly and are apparently marked for continued rapid advance.

[12] The term "identification" is used here in the sense in which it is employed by H. A. Simon, *Administrative Behavior* (New York: The Macmillan Company, 1947, 1957), p. 205: "A person identifies himself with a group when, in making a decision, he evaluates the several alternatives of choice in terms of their consequences for the specified group."

members of the Forest Service with the well-being of the organization, linking their own positions and welfare and futures with those of the agency, fusing their perspectives with those of their colleagues and superiors. It is a set of conditions that sets them apart from all people "outside" the organization, binds them intimately with other organization members; that "injects into the very nervous systems of the organization members the criteria of decision that the organization wishes to employ," and thereby vastly increases the probability that each of them will "make decisions, by himself, as the organization would like him to decide."[13] Without realizing it, members of the Forest Service thus "internalize" the perceptions, values, and premises of action that prevail in the bureau; unconsciously, very often, they tend to act in the agency-prescribed fashion because that is the way that has become natural to them. Much of what the Service does tends to further this process.

TRANSFER AND PROMOTION

For example, transfer of personnel is treated in the Forest Service as a device for "the development, adjustment and broadening of personnel"; consequently, men are deliberately moved a good deal, particularly during their early years in the agency. The Service does not merely wait until vacancies occur; it shifts men to replace each other in what looks like a vast game of musical chairs, but for the serious purpose of giving them a wide range of experience in preparation for advancement to positions that require a broader understanding of national forest administration than can possibly be gained in long assignments at a single duty station. If transfers can be coupled with promotions, the added incentive to move is provided; however, "horizontal transfers [i.e., in the same grade] also may be proposed as

[13] *Ibid.*, p. 103.

a prerequisite to possible future advancement." If an individual declines a proposed transfer, his status in his old post is not prejudiced, for transfers are recommended, not ordered formally, in most cases,[14] and the *Manual* decrees, "There will be a clear-cut determination that the transfer will not work undue hardship either on the transferee in his personal situation or on the receiving or sending unit"; nevertheless, the *Manual* also warns the Service "is forced to insist that those who wish to advance must, at times, waive personal preferences as to location, make inconvenient moves, and serve where most needed . . ." Hence, when most men are asked to move, they move; it is chiefly old hands with long years of service, no longer interested in rising any higher than they are, who furnish the few declinations. Younger men just starting their careers rarely do, for such an action might impair their futures.

The Rangers studied here have had differing experiences, but all have employment records that reflect the general statements of transfer policy. Of three with more than 20 years in the Forest Service, one was in five locations, one in four, and the third in three within a dozen years; and each moved again at least once later on. As for the younger men, one has served in four places in five years, the other in two places in seven years. Three served in one capacity or another on the staffs of forest supervisors as well as at the district level.

In ten or fifteen years, then, a man in the Forest Service is introduced to many of the problems and practices of national forest administration; he is doubtless "developed, adjusted, and broadened." But the impact of rapid transfer is more profound than training alone; it also builds identi-

[14] One forest supervisor reported he thought he might have been brought up on charges of insubordination if he refused to move. Actually, this seems most unlikely, but this expression of concern by a high-ranking officer suggests how much importance is attached to transfer.

fications with the Forest Service as a whole. For during each man's early years, he never has time to sink roots in the communities in which he sojourns so briefly. He gets to know the local people who do the manual work in the woods, but not very well in the short time he spends with them. He barely becomes familiar with an area before he is moved again. Only one thing gives any continuity, any structure, to his otherwise fluid world: the Service. When he reports to a new area, his superior helps him get installed in his new living quarters, introduces him to the townsfolk who will be his neighbors, acquaints him with all the members of the local work force and fire organization, instructs him in the management of the administrative unit, supervises and evaluates and corrects his work, and prepares him to shoulder heavier responsibilities. Whenever a younger man severs his ties in a location to which he has just become adjusted and takes a new place, an experienced Forest Service officer is there to receive him, support him, guide him. To be sure, there are strains and conflicts and frictions, too. But, in general, it appears from the limited evidence of this study that the men who move rapidly are received sympathetically by those to whom they are detailed, and are taken in hand for a time as the personal as well as the official responsibilities of their immediate supervisors.[15]

Thus, the Forest Service acquires a more or less fraternal aura for its newer members. To be sure, it is the organization that uproots and shifts them in the first place, but the hardships are considerably softened by the visible team of friends and colleagues ready to help them and to make the

[15] And, at times, of their subordinates. One nonprofessional employee who had served the Forest Service for almost thirty years on one district worked under a total of thirteen Rangers in that period. He took a paternal interest in the young men who supervised him, and the Ranger under whom he was serving at the time of the interviews for this study confessed his great dependence on the older man. "My chief job," the employee said, "is breaking in Rangers," a job he apparently performed proudly, gently, and with affection for those he helped.

transitions as smooth and pleasant as possible. Moreover, behind the inconveniences stands the comforting knowledge that transfer is preparation for advancement, that every assignment and detail is recorded, adding to qualifications for promotion when the opportunities arise.[16] The impersonality of the system is reduced, the sense of belonging enhanced—particularly since the frequency of transfers to different locations and administrative levels brings many of the men in each region into personal contact with each other despite the dispersion of the agency.[17] Everywhere, they encounter men with similar interests, similar problems, similar objectives, similar aspirations, similar complaints. They find understanding and appreciation of their problems. Their ties with their fellow-officials are multiplied and deepened. As they become part of the organization, the organization also becomes part of them.

The opportunity for this process of organizational acculturation to have a chance to work its effects on every executive in the Forest Service (i.e., every officer at the Ranger level or higher) is maximized by two practices: One is a firm system of promotion from within for professional positions in national forest administration. The other is a policy of relatively unhurried promotion.

The Forest Service, at least as far as its professional foresters are concerned, constitutes a classic illustration of a career system. The foresters are comparatively safe from the vicissitudes of politics and economics; the merit system has

[16] In many organizations, transfer is employed as a sanction, too. Choice assignments go to the men who fit the organization patterns, undesirable ones to those who depart from the preferred patterns, and to newcomers. (The policeman detailed to the "sticks" is a familiar case in point.) No evidence of this strategy was found in the Forest Service in the course of this study, nor did any of the men interviewed at any level seem to consider this a likely penalty. If it is used at all, it is apparently used sparingly. (However, see footnote 14 above.)

[17] Inter-regional transfers are not uncommon, but they are far less frequent than intra-regional shifts.

guarded them effectively since the first days of the agency, and reductions in appropriations are absorbed more by the large seasonal work force than by permanent officers. With professional staff thus stabilized, the Service has been able to recruit its new professionals to fill entering, sub-Ranger positions, and to fill virtually every job at the Ranger level and above by advancing someone from a lower grade. Says the *Manual:*

> In filling a position, we should avoid or minimize the chance of bringing into the Forest Service persons about whom we know less, because of lack of service and salary records, than we do of our personnel and who might therefore be incorrectly appraised by us as being better qualified than our own personnel.

Without exception, all the Rangers interviewed, though professional foresters, served their "apprenticeships" in sub-Ranger grades; in fact, there are no Rangers now in the Service recruited directly from outside the Service. Furthermore, almost all the officers higher than Ranger in national forest administration have served as Rangers at some time in their careers. To be sure, there is no single set pattern for advancement; there are many alternative ladders, many different routes upward. But they all have one thing in common: vacancies are filled by promotion and transfer, never (for all practical purposes) by lateral entry.[18] There are no "strangers" in the administrative positions in national forest administration, save for the men junior to the Rangers. There are compelling reasons for this practice; it is a long step indeed from forestry school to the heavy responsibilities of district management. Nevertheless, it means the ethos of the agency is subject to few jarring dissonances from within.

[18] Indeed, every Chief since Gifford Pinchot has come from inside the Service with but one exception, and even he had had extensive Forest Service experience when he was brought back under the New Deal to reorganize the bureau.

Not only are the higher positions filled by men selected from inside the Service, but men in the lower positions ordinarily occupy them long enough for the process of acculturation to take effect. In part, this is a result of circumstances beyond Forest Service control rather than planning; civil service law requires that employees remain in grade for a year before being promoted, and, in any event, the pyramidal structure of the organization for national forest administration provides fewer openings for advancement than there are men eligible to move ahead. In part, however it is also deliberate policy; it is estimated by the Service that it will take not less than three years to rise to the command of a Ranger district, and possibly as long as a decade. (One of the Rangers interviewed traversed this distance in less than the minimum, two took three years, one took eight years, and one was twelve in coming up.) The minimum for becoming a forest supervisor is seven or eight years, while seventeen years is a short time to ascend the ladder to regional forester; generally, most men in these positions took longer.[19]

As private industry absorbs ever larger percentages of technically trained foresters, and as intensification of forest management creates additional positions in the Forest Service, the rate of advancement tends to become more rapid. Nonetheless, forest officers are exposed for substantial periods to the environment of the agency before they are appointed to executive positions, and they remain always under the surveillance of men who have spent practically their whole adult lives in Forest Service employ. They are absorbed into the organization by a kind of gradual social osmosis, during which they, in turn, absorb many of the

[19] "The average grade 9 Forester is 41 years old and reached his present grade in 8 years. For GS-11 the same statistic is 44 years of age and 15 years of service to reach it; for GS-12, the age is 48 and elapsed time from entry to present grade is 22 years; and for GS-13, age 50, elapsed time is 23 years." W. A. Elkins, *op. cit.*, p. 5.

prevailing values, assumptions, and customary modes of operation.

Seniority does not automatically bring promotion. Periodically, the regional foresters and assistant regional foresters in each region assemble to review the records of the men under them and determine what their assignments should be. Length of service is a factor in their judgments, to be sure, but it is far outweighed by other elements—principally the evaluation of each man by his immediate superior, by the personnel management division of the regional office, and by the regional forester and his staff (most of whom will have met almost all the professional foresters in their jurisdiction in the course of visits, conferences, inspections, training meetings, and conventions of professional societies and associations, and who also have access to inspection reports, civil service performance ratings, and other appraisals of accomplishment and potential). Some men will be transferred to round out their experience; others will be left where they are to season; still others will be advanced; a number are repeatedly passed over. Some never get beyond the Ranger level, and a few serve out their years without even achieving command of their own districts; some shoot comparatively meteorically through the hierarchy, and many a veteran Ranger has been inspected by a man the Ranger himself trained a few years earlier.

It does not take most men long to learn there are attributes rewarded by the organization, and those who yearn to rise deliberately cultivate those attributes if they can. There is a striving on the part of many to demonstrate they fit into the approved pattern, and even those who profess to be indifferent to promotion cannot help picking up many of the traits of the culture in which they work. For those who fit the pattern naturally, this takes no conscious effort. For others, it takes conscious self-appraisal and adjustment; as one Ranger who failed to advance from his sub-Ranger assignments during his early years put it, "I changed my outlook and reorganized myself and my own line of thinking," and the transformation was followed by promotion. In other words, promotion is a sanction, a reward or a punishment based on excellence on the job, but excellence on the job is in practice measured by the proven predisposition to behave in the organizationally desired fashion rather than just by technical proficiency.[20]

Promotion and transfer are thus far more than methods of staffing. As practiced in the Forest Service, they also foster in each officer identification with the agency—with its survival and welfare, with its goals, with its procedures, with its members.

THE USE OF SYMBOLS

Identification is heightened by the use of symbols. Perhaps the outstanding ones are the uniform and the badge.

[20] One regional forester commented: "We would simply like to add a thought or two [here] which we believe has considerable to do with a personnel placement. The first one is that nearly all Forest Service personnel agree that Ranger and Supervisor jobs are the most desirable in the Service because of the great personal satisfaction. In these jobs, the incumbents are line officers in charge of a unit area and can see the accomplishments much more concretely. The second thought is that many of the personnel through their own personal desires to remain Rangers or Supervisors often have considerable affect on what their careers become." Some men, then, prefer to remain in the field rather than advance to positions more remote from the woods. However, from the writer's discussions with forest officers, this would not seem to be the general attitude among men in the earlier stages of their career. Moreover, it appears likely that those who elect to stay in the field do not find onerous the patterns of behavior they are required to observe. Thirdly, as noted above, even those unconcerned about promotion unconsciously adopt the outlook and characteristics of the organizational culture in which they are immersed. Finally, Rangers who demonstrate they cannot or will not observe the requirements of the organization *may* be shifted to positions of less discretion and responsibility. Hence, the preference of some men for field assignments does not by any means nullify the process of organizational acculturation described here.

The whole purpose of uniforms and badges is to identify the members of organizations, to differentiate the wearers from everyone else and to link them with each other. The livery and insignia show at a glance who is "in" an agency and who is not, and establish authority and status at a glance. And they also foster a group spirit and unity, a "we" feeling, a common bond.

In fact, the attitude of the men in the Forest Service toward the uniform varies. Regulations make the wearing of it mandatory in smaller towns, optional for men stationed regularly in larger cities. It is required for appearances in public capacity as spokesmen and representatives of the agency. There is an allowance for the purchase of uniforms. However, some men prefer to wear work clothes most of the time—particularly when dealing with loggers and grazers, before whom they prefer to appear as individuals doing business rather than as authoritative agents of a government bureau—and are regularly admonished by their superiors to get into their "greens."[21] Some wear the uniform, with pride, whenever they can. Many wear at least part of it most of the time. Yet, although the reactions are mixed, and the observation of the rules somewhat spotty, it is significant that a majority of the officers in the Forest Service expressed a preference for retaining the uniform when polled on the question a few years ago. Despite the unwillingness of many of them in some regions to wear it as regularly as directed, it is a symbol most are not ready to relinquish; the privilege of wearing it still unites them.

[21] For example, in a report on a General Integrating Inspection of one Ranger district, a forest supervisor wrote in 1957: "Another item believed worthy of comment is the use of the uniform. With an Assistant Ranger position permanently assigned to the District, more stress is needed on the proper use of the uniform. During the past ten years or so many of the field personnel have gone to 'work clothes' and have neglected to wear the 'greens.' This situation should be corrected in accordance with manual guide lines, and the Assistant Rangers in particular should have the proper example set by their immediate superiors."

The Forest Service insigne—the shield-shaped badge with the agency name and a tree emblazoned on it—is a familiar and respected one the country over. In Washington, the agency uses distinctive wooden plaques rather than the standard signs to identify its offices, while rustic signs bearing its emblem appear on almost all the properties it manages. Indeed, it has been said the adoption of the designation "Service"—now a fairly commonplace term, but a novelty when it was originally selected—instead of the more common "bureau" helped set it in a class by itself, accentuating its self-consciousness and corporate spirit.[22] These are all small things, but they do set the agency apart. Many public servants, asked who their employer is, are likely to name "the government," or perhaps their department. Forest officers will almost invariably respond, "the Forest Service."

The agency symbols, even when they are not enthusiastically supported, keep the members aware of their membership, and encourage them to think in terms of the agency. Consequently, in the course of time, its premises tend to become their own.

HEADQUARTERS CONSULTATION WITH FIELD OFFICERS

Identification is further intensified by the Forest Service practice of sounding out field opinion on questions affecting field administration. Social psychologists have indicated that participation in the formulation of organization decisions tends to promote identification.[23] The Forest Service provides

[22] See J. M. Gaus and L. O. Wolcott, *Public Administration and the Department of Agriculture* (Chicago: Public Administration Service, 1940), pp. 265-66.

[23] See, for example, G. W. Allport, "The Psychology of Identification," in S. D. Hoslett (ed.), *Human Factors in Management* (New York: Harper & Bros., 1946, First Edition): "We are learning some of the conditions in which reactivity [i.e., "rebellion against authority, . . . disaffection of all sorts"] does decline. . . . Opportunities for consultation on personal problems are, somewhat surprisingly, found to be important. And as members of S.P.S.S.I. [Society for the Psycho-

many opportunities and channels for such participation.

For instance, the field is sometimes polled formally. As noted earlier, the Washington office requested opinions from the field on whether uniforms should be required, and also on the style of the uniform; the attitudes of members of the Service down to the Ranger district level were obtained, and the policy eventually adopted was based in large part on their reactions. In another case, the Washington staff, anticipating legislation on overtime pay, proposed a tentative stand for the Service to present to the Secretary of Agriculture and Congress; the circular was transmitted to the field with the comment, "As usual we wish, of course, to check with you and have your advice on these—and any other related problems—before making specific recommendations to higher authority." (Reports from the regional foresters on opinion in their jurisdictions occasionally indicated that the regional foresters did not agree with the judgments of their subordinates, but they dutifully transmitted those judgments just the same. A couple of regional foresters reported comments obliquely critical of the Washington office on the overtime issue, urging headquarters to simplify and stabilize policy.) In a third instance, suggestions for administrative improvement were solicited from the field by Washington; the responses were analyzed, and

logical Study of Social Issues] have shown, group decision, open discussion, and the retraining of leaders in accordance with democratic standards yield remarkable results. One of Lewin's discoveries in this connection is especially revealing. People who dislike a certain food are resistant to pressure put upon them in the form of persuasion and request; but when the individual himself as a member of a group votes, after discussion, to alter his food habits, his eagerness to reach his goal is independent of his personal like or dislike. In other words, a person ceases to be reactive and contrary in respect to a desirable course of conduct only when he himself has had a hand in declaring such a course of conduct to be desirable." (P. 259) See also M. Sherif and H. Cantril, *The Psychology of Ego-Involvements* (New York: John Wiley & Sons, 1947), pp. 369-71.

organized into over a thousand separate recommendations, some minor, others sweeping. Priorities were assigned to the recommendations according to their urgency and the number of times they were mentioned, and projects to put them into practice were instituted. Almost all the proposals, suggestions, and complaints have been acted upon, and work continues on those not yet effectuated. Such massive surveys do not occur every day, but they are made frequently enough, and are supplemented by similar inquiries from the regional offices, to betoken to field men that they are not simply passive instruments manipulated by the agency leaders. Some agency decisions are clearly choices in which they have had a part, even when their personal preferences do not happen to prevail.

Over and above the polling techniques, the Forest Service provides many additional opportunities for field men to make their views known to their superiors. Indeed, these may be even more effective means of encouraging an upward flow of ideas and opinions from the field, for they do not restrict the comments to particular subjects. Thus, as previously noted, few men who are under inspection pass up the chance to ventilate their suggestions and criticisms—both during working hours, and even more so in the informal social hours after the close of business; this practice is not only tolerated, but encouraged, and members of the Service insist that their recommendations and complaints *do* get back to the higher levels, sometimes generate action, and do not (unless carried to an extreme) result in injury to the sources for being outspoken. Since Rangers are inspected by supervisors' offices, regional offices, and from time to time as sample districts in their regions by the Washington office, this often permits them to air their convictions quite effectively. The opportunity *can* be abused; the chronic complainer, the "whiner," and the destructive critic are apt to be heavily discounted. But most men are conscious of the possible abuses, and they are able to inject their predilections into the decision-making process with some hope of affecting the final product. It is apparently not just an empty ritual, for Ranger complaints (and complaints from other levels) about excessive paper work and the unwieldiness of the *Forest Service Manual* are generally regarded as the principal factors behind projects to reduce the former and simplify the latter to which the Washington office has devoted much time, energy, and money. In any case, the practice contributes to the general sense of participation. So, too, does the standing official invitation to "inspectees" to file objections to any aspects of inspection reports they consider unfair or inaccurate; many of the men interviewed have registered protests at one time or another, and some have been sustained in their protests.[24]

Forest supervisors and regional offices often seek the advice of Rangers on pending questions quite informally, apart from Service-wide polls and regular inspections. For example, Rangers have been asked about the relative merits of dividing districts as compared with adding to existing staffs and maintaining existing boundaries. If division is chosen, they are consulted about the lines of the new dis-

[24] The members of the Service are invited to regard inspections as a co-operative rather than an investigatory procedure; they are asked to consider inspections as a means of participating, asked to do so for the explicit reason that this builds identifications. "This attitude towards sharing responsibility," said the Chief in a letter (February 21, 1955) to an Assistant Secretary of Agriculture, "naturally makes each man want to help the man with whom he does the sharing. This mutual-help attitude is reflected in the character and tone of inspections. We always try to recognize good work as well as to point out needed improvement. . . . The attitude is to see how together we can do a better job. I explain this partly because many people do not understand why we customarily state in inspection reports that something 'should' be done rather than say it 'shall' be done. I suppose that with us, should means shall, but we try to avoid the master-and-slave attitude that tends to weaken individual initiative and a sense of joint responsibility." The contents of the letter were reproduced, and circulated in many parts of the Forest Service.

tricts, and about the best locations for the new Ranger stations. When projects for national forests are administered from supervisors' offices, the Rangers are drawn into the planning; roads, recreation areas, and other improvements are installed only in consultation with them. A couple of Rangers reported stopping work being done by project crews working out of higher headquarters because the Rangers disapproved of the way the work was being done. In an earlier chapter, mention was made of abandonment by a forest supervisor of plans to spray an area because the Ranger thought the public reaction of his district would be strongly unfavorable. When adoption of new equipment or practices is contemplated, they are tried out on pilot Ranger districts, and the evaluations by the Rangers play a large part in the decisions to adopt, modify, or reject the machinery or procedures. It would be an exaggeration to say that the Rangers are consulted about *every* decision affecting the management of their districts, but there can be no question that consultation on many matters of concern to them is common enough to lend credibility to the impression that the Rangers participate actively in the formation of administrative policy for national forest administration. Nor do they necessarily wait to be asked for their opinions; they not infrequently take the initiative and urge their ideas upon their supervisors and other superiors.

Conferences for budgetary, policy, or training purposes afford additional opportunities for field officers to impress their concepts on their superiors. When Rangers assemble to work out financial plans with their supervisors and the supervisors' staffs, they come with proposals of their own with regard to their programs for the fiscal period, and they bargain with each other and with their superiors to get as close to their objectives as they can. When district programs are coordinated to mesh with forest policy, forest policy is often adjusted to fit district needs as the Rangers see them.

At training conference discussions and seminars, Rangers do not hesitate to point out defects in prevailing policies and to suggest the remedies they favor. Theoretically, all of this could be suppressed. Fiscal and program plans could be arbitrarily decided by the higher offices and simply transmitted to the field for execution. Training conferences could be confined to discussions of how promulgated decisions should be implemented rather than being permitted to range over assessments of the decisions. But this is not the way the Forest Service operates. In part, the leaders may invite participation by field men to ensure the practicability of the decisions the agency reaches; in part they may do it to ensure a minimum of opposition from the field. Whatever the motives behind them, the conferences furnish field men with avenues of access to their leaders, and the field men are encouraged to use them. And they do.

Furthermore, the Rangers are told over and over again that they are the pillars on which the Forest Service rests. According to the *Forest Service Manual* itself:

> The Forest Service is dedicated to the principle that resource management begins—and belongs—on the ground. It is logical, therefore, that the ranger district constitutes the backbone of the organization.

And the Chief of the agency, in a speech to the 1958 convention of the National Woolgrowers Association (who have often been critical of Forest Service grazing policy), told the assembled sheep raisers:

> The man who is responsible for making the initial decisions for the management of your individual allotments [of grazing privileges] is the district ranger. He lives in your community as a neighbor; his children go to the same schools as yours do. You can be sure that he would not propose livestock reductions that sometimes lead to bitter controversy if he were not thoroughly convinced this action is necessary . . .
>
> We have often heard it said that the rangers and supervisors are good guys but that they are merely doing what they are

> told to do by some bureaucrat in Washington. It would be utterly impossible for the small staff we have in Washington to be sufficiently familiar with conditions on all of the national forests to make or even to suggest what specific decisions should be made as to the management of individual allotments. Of necessity we have had to delegate responsibility and authority to the men on the ground. We, in Washington, establish general policies and procedures. We make periodic checks and inspections to determine how well the policies and procedures are being carried out, but the responsibility for making the decisions and the authority to carry them out has been delegated to the men on the forests and ranger districts.

Lest it be thought he was trying to evade his own responsibilities, the Chief added:

> In the final analysis I am responsible for the action of all the members of the Forest Service. Although we delegate responsibility and authority all the way down the line, I cannot shift the responsibility for the work of the Forest Service to anyone else's shoulders. You might say I 'share' my responsibilities with the regional foresters, forest supervisors, and rangers, but in sharing it I do not escape any of the responsibility for what happens.

Skeptics might argue the speech was a smokescreen, although it is not so regarded in the Service. But even a skeptic cannot help but be impressed with the fact that the field men are so visible and respected in their communities that the Chief himself—whether engaged in a maneuver to relieve his office of pressure or sincerely depicting the realities of decision-making in his bureau—sometimes takes refuge behind them. It is persuasive evidence that they make important decisions in the Forest Service, and play significant roles in the administration of the agency. The speech was circulated to all the members of the Service on national forests in grazing regions.

Actually, the field men do not seem to need convincing on this point. If anything, it would appear difficult to *alter* their convictions about it. For their day-to-day experience has already persuaded them that all higher headquarters are heavily dependent on them—not only for executing policy pronouncements, but in formulating them as well. Again and again, the researcher is told by officers in the field that they do the bulk of the work even though others sign the papers, and their superiors freely acknowledge this dependency. True, Rangers can complete only a few transactions, and must submit most to higher offices for completion; true, their work is reviewed, and their suggestions often modified or rejected; true, they are inspected, and inspectors check in the woods as well as in the office; true, they are subject to a battery of controls from above. But they know their districts more intimately than any of their superiors. They do the leg work on which Service assessments of the capacity of the land are based. They draw up the plans from which production quotas and targets are derived. They furnish the data for statistical analyses. They plan sales, and make findings and recommendations on which issuance or denial of permits depends. They do the fundamental work for land exchange, though final action depends on a Cabinet-level commission. In short, leadership decisions about what the Forest Service can and should do rest in the last analysis on what the field men tell the leaders. Even the workload calculations so central to budgeting are computed from observations made on sample Ranger districts, and evaluations of equipment come from Ranger experience with each piece of apparatus. The factual premises on which policy decisions are based are furnished in large measure by field officers. They do not have to be told this is the case; they believe it already. Whether or not the Rangers magnify their influence—and it is by no means clear that they do—their belief in their influence is apparently genuine and widespread. That Rangers participate actively and significantly in the running of the Forest Service is taken for granted.

It does not seem likely that all the types and evidences of

field participation in agency decisions were deliberately instituted or adduced to create the feeling of identification with job and organization that social psychologists say is linked with a sense of participation. Rather, the various kinds of consultation probably grew out of the nature of resource management problems. But it makes little difference for this study whether the practice was designed to enhance the sense of participation or was simply the incidental fruit of the pursuit of other objectives. The fact is, field officers *do* participate, and to a degree they seem to believe is significant. Thus, they come to identify themselves with the Forest Service and its decisions.

THE FIELD MAN IN THE COMMUNITY

People's concepts of themselves and of their places in society reflect to a large extent the way others act toward them and react to them. Each individual's image of himself—his picture of who he is, what he is, where he belongs, how he should behave—is determined partly by the way it appears to him *others* picture him. How others see him, and what they expect, is indicated by cues from the environment:

> An individual identifies himself with and regards as a part of himself the particular constellation of values he learns from his environment. On the basis of this learning the individual defines his own role or status: he learns what group or groups *he* belongs to; what other groups are regarded as "higher" or "lower" than his own; what groups are to be regarded as enemies, antagonists, or competitors; what as allies, helpers, or friends.[25]

The cues that come to Rangers from their environments tend to identify them with the Forest Service. Many of these, of course, come from their superiors, coequals, and sub-

[25] M. Sherif and H. Cantril, *op. cit.*, p. 135. See also, R. Linton, *The Study of Man* (New York: D. Appleton-Century Co., 1936), Chapter VIII.

ordinates in the organization. Many, however, come from the people outside—from their friends and neighbors, their business associates, their clienteles. When a Ranger takes over a new district, he is generally invited promptly to join local civic and community organizations—partly because his position as manager of large properties automatically makes him a person of some standing in most localities, partly because the Forest Service is always "represented" in such associations. It is with the Rangers that loggers, ranchers, picnickers, and permittees of all kinds do business—both in negotiating agreements with the Forest Service, and when the agreements are supervised. The Rangers are therefore shown considerable deference. The Rangers are cast in the role of law-enforcement officers when trespasses occur; to violators, they often appear, and are treated, as figures of authority. Men engaged for emergency fire fighting see them as fire bosses in full charge of complicated and dangerous operations. They appear before school and college groups, associations of young people (4-H Clubs, Future Farmers of America, etc.), garden clubs, hunting and fishing clubs, and similar groups in fulfillment of their information and education responsibilities (especially for fire prevention purposes). To many local residents, they are employers who provide seasonal employment. In business circles, they appear as executives managing tens—even hundreds—of thousands of acres of valuable land worth millions of dollars and doing thousands of dollars worth of business every year. For most people, in short, they stand for the Forest Service; indeed, they personify the Forest Service. The role is thrust upon them.

Since the way men think of themselves is shaped partly by the way others demonstrate they think of them, and since Rangers are reminded over and over again that they are viewed as representatives and spokesmen of the Forest Service, it stands to reason that they come to identify themselves with the Forest Service. Not only is their role defined for them by the agency; the definition is reinforced by the community. Even the tendency of local populaces to emphasize the personal rather than the official attributes of the Rangers as they are assimilated into localities is counterbalanced to a degree by the rapidity of transfer; from the point of view of the people in a given area, individuals come and go, but there is always the district Ranger of the Forest Service. Eventually, this is how the Rangers see themselves—as parts of the organization rather than simply as individuals.

PUBLIC RELATIONS

In addition to the information and education activities of forest officers in the field, the Forest Service conducts a systematic public relations campaign from its Washington and regional offices. Speakers are sent to meetings, films and film strips are prepared and made available to interested groups, publicity releases are prepared for the mass media of communication, public reports are widely distributed, pamphlets and leaflets about the Service are printed by the Government Printing Office and distributed by the Department of Agriculture and the Superintendent of Documents as part of the government's general information program. The National Advertising Council, as a public service, cooperates with the Forest Service in pressing a campaign against forest fires (using "Smokey Bear" as a symbol). The public relations effort is on a large scale, systematic, and continuous.

Actually, the principal purposes of the public relations program are to acquaint the public with the work of the Forest Service, to make the public fire-conscious and thereby to reduce the incidence of man-caused fires, and, to a lesser extent, to facilitate recruiting of personnel. But the effects are more far-reaching. In the first place, it counters the propaganda of Forest Service critics; by presenting its case as widely as possible, the Service builds grass-roots support for its policies, or, at least, neutralizes the opposition that might develop if the attacks went unanswered. Generally, Congressional attitudes more or less reflect their constituencies, attitudes; this means that Congressmen under pressure from interest groups seeking heavier use of the national forests than the Forest Service deems safe are subjected to counterpressures from the friends the agency has won in their constituencies, and that Congressmen from industrial states having few or no national forests (and who might therefore be indifferent to the forestry program) will often display a friendly concern for the Service. Moreover, the public-relations materials acquaint many users of the national forests with the objectives of the Service, and may reduce somewhat the opposition with which field men must cope.

In the second place, these materials are sent to the members of the agency, providing them with arguments for its programs and policies, reminding them of its purposes and methods, keeping them informed of its problems and the approved solutions, offsetting the effects of criticism and hostile reporting, and reinforcing their dedication to the Service and its activities. The public-relations campaign also creates both inside and outside of the Forest Service an image of the Service as the uncompromising champion of the public interest and welfare, as a defender of public property against spoilation by powerful but selfish special interests, an image that builds up the pride of men associated with it and elevates their personal prestige along with its own.

Public relations is designed to reduce resistance from outside the Service, win support when possible, and counteract centrifugal tendencies that might induce field men to deviate from promulgated policy. It is intended to affect the external forces acting *on* the Rangers while it strengthens "*inside*" them, by heightening their identification with the organization, tendencies toward conforming with agency decisions. It is not always and everywhere completely successful,[26] but it seems to have been quite effective generally.

THE INTERNALIZATION OF FOREST SERVICE PERCEPTIONS, VALUES, AND PREMISES OF ACTION

Much that happens to a professional forester in the Forest Service thus tends to tighten the links binding him to the organization. His experiences and his environment gradually infuse into him a view of the world and a hierarchy of preferences coinciding with those of his colleagues. They tie him to his fellows, to the agency. They engender a "militant and corporate spirit," an organized "self-consciousness,"[27] dedication to the organization and its objectives,[28] and a fierce pride in the Service. They practically merge the individual's identity with the identity of the organization; the organization is as much a part of the members as they are of it. At least some of the practices described above were probably initiated with this in mind, but a number were apparently adopted for other reasons and contributed to this result more or less accidentally. Still, whatever the purposes, one outcome of the practices is that field officers (among others) make their administrative decisions in terms of the consequences for the Forest Service, and in terms of criteria the leaders of the Forest Service wish them to employ.

[26] According to some observers, there have been "boomerang" effects. Field men who accepted the idealized image of the Forest Service were disillusioned to discover that the leadership sometimes found discretion the better part of valor and adopted strategies calculated to placate some politically powerful users of the national forests.

[27] J. M. Gaus and L. O. Wolcott, *loc. cit.*

[28] "Many Forest Service men are dedicated in the sense that they will fight for certain courses of action, and against others, almost without regard to the cost in terms of wear and tear upon their own emotional systems or in terms of public criticism and opposition." From a letter to the author from M. Clawson, former head of the Bureau of Land Management.

The Result: Voluntary Conformity

Forest officers are selected in a fashion that winnows out many of the men who probably lack the inherent predisposition to conform to the preformed decisions of the Forest Service, and that guarantees at least a minimum level of technical competence. Their competence is broadened and deepened by post-entry training, both in-service and outside, and by placement, transfer, and promotion policies; the methods of improving technical skill also intensify the predisposition to conform. The predisposition is strengthened by generating identification with the agency (which at the same time adds to understanding of the announced agency objectives). As a consequence, officers of the Forest Service conform to agency decisions not simply because they have to, but because they want to. And they can because they have been equipped to do so.

"Wanting" to conform is used here not to mean an abstract desire to be obedient the way a child wants to be "good" but not to do any of the things that being good means to his parents. Rather, it is employed to mean wishing to do as a matter of personal preference the things that happen to be required. It is in *this* sense Forest Service personnel want to conform. Often, confronted by a situation in the field, there is a course of action they would "instinctively" like to follow, that seems "clearly" to be the "best" and "proper" one; a good deal of the time, this "happens" to be the action prescribed by the Service. That is, they are not consciously "conforming"; they are merely doing what is "right."[29]

[29] For example, in one region a number of years ago, travel and expense allowances were scaled according to the grades of the personnel involved. Thus, though a superior and a subordinate working together away from their homes might be in close association, the former received a larger reimbursement. Asked whether it might not be reasonable to eliminate, or at least reduce, the rather substantial differential, a subordinate's reply was emphatically negative: "Why, that wouldn't

Inevitably, there is little consciousness on their part of the deliberate search for the appropriate "instincts" by the leaders of the Forest Service, and of the deliberate efforts to cultivate these and weed out others.

Indeed, even when men are overruled—when their "instincts" do not move them in the same directions as their superiors—their adherence to the provisions of the higher decision comes in a sense from inside themselves. They rarely persist in opposition after judgment has been rendered, or engage in administrative sabotage, or carry appeals to higher levels. In part, of course, this obedience is based on the risks of such action.[30] In addition, however, it rests on the widely expressed sentiment that "there's no other way to run a big organization." They value the organization more than they value getting their own way; they therefore carry out directives they opposed, because doing so is "necessary" and "right"—and, though they do not seem to be aware of it, because this feeling is carefully nurtured by the organization.

The thrust toward behaving in accordance with the preformed decisions issued by the Forest Service is thus not imposed on a reluctant and resistant body of men; it is as

be right," he said. "When I go out cruising timber, say, and I have to spend the night out, I can go to a tourist cabin or a cheap hotel to get a place to sleep, and I can eat in local diners. But the supervisor can't do that. He has to keep up a front. He meets important people, and he can't get along the way I do. Why even the Ranger who works out in the woods with us a lot of the time can't always do the way the rest of us fellows do; it costs him more to do things the way he has to, so why shouldn't he get more?" The fact that other agencies did not draw a distinction by grade did not impress him; the differential seemed "proper" to him and offended neither his sense of abstract equalitarian justice nor of material self-interest. (There were complaints about linking the differential to rank rather than expenses, but few about the appropriateness of some kind of differential. Eventually, however, the differential was abolished for administrative reasons.)

[30] Other factors in obedience are discussed in H. A. Simon, D. W. Smithburg, and V. A. Thompson, *Public Administration* (New York: Alfred A. Knopf, 1950), pp. 188-201.

much internal as external. But the internal forces are not left wholly to chance; although some have developed incidentally, they are also encouraged and even planted.

This completes the inventory of methods by which the Forest Service manages to integrate the organization for national forest administration in spite of the awesome tendencies toward fragmentation to which that organization is heir. But, in order to make the presentation wieldy and clear, the categories of influences on the administrative behavior of district Rangers have been treated as though they were independent of each other. In fact, of course, they cannot be isolated from one another. The interplay among the influences—among the centrifugal and the integrative, the external and the designed and the fortuitous—is the subject of the next, and concluding, chapter. Here, the elements of the organization, heretofore artificially separated for examination, are reassembled and portrayed in their "natural" state.

CONCLUSIONS

CHAPTER VII

ATTAINMENTS AND DILEMMAS

The Conquest of Centrifugal Tendencies

If unity and uniformity were the same thing, it would be relatively easy to determine just how unified an organization is. However, when an organization functions through a field network under a wide variety of conditions, so that different decisions and different actions in scattered units may be fully consistent with each other and with a common policy statement, conclusions about unity are much harder to reach. In the last analysis, they must be based on circumstantial evidence and impressions.

On this basis, the Forest Service organization for the administration of the national forests must be reckoned as having achieved a high degree of unity. The measures are not precise, but they all point to this conclusion.

In the first place, over-all performance comes remarkably close to the goals set by the leadership. There are timber sales targets and timber cut targets; maximum acceptable burn targets for controlling fires; grazing targets; wildlife targets; revenue targets; targets for the number of visitors to be accommodated; and many others. Over the years, the averages of actual performance figures fall within a few percentage points of the objectives. What is more, when the leaders of the Forest Service set long-range objectives, and Congress provides money for the activities, output rises in accordance with their plans. Conceivably, the leaders could be following rather than leading their field men; they could be simply predicting what the Rangers will do, as the collective production of the country's individual factories is forecast, but not controlled. Since the goals of the leaders are formulated on the basis of field reports and estimates, and of field accomplishments, this is certainly a logical possibility. In fact, however, three things appear too pronounced, particularly when compared with the private sector of the forest economy, to be the result of unco-ordinated actions of individual Rangers: the consistent, long-term connection between announced goals and actual performance; the responsiveness of production to changes in leadership objectives; and the steady march of performance records toward goals of the leaders (from the early days of the agency to the present day, and, if the past is at all indicative of the future, into the future). Despite the centrifugal forces at work in national forest administration, the actual accomplishments of field units have been brought into agreement with the mission defined by the officials in central headquarters.

In the second place, one seldom finds the symptoms that often accompany organizations in which field officers have ceased to respond to unified policy direction—almost no charges of administrative sabotage by frustrated leaders, for example; comparatively few accusations of local favoritism and discrimination by the clientele of the national forests; no discoveries by Congressional investigators of scandalous field collusion with special interests (like those that produced sweeping reorganizations of the Internal Revenue Service); little of the internal warfare of the kind that has rent the Defense Department recently, and earlier beset the agencies established for administration in the Second World War.

True, the pathology of an organization must usually reach a rather advanced state before it regularly exhibits such grave symptoms. But they appear intermittently even in organizations ordinarily considered "healthy," and their infrequency in the Forest Service, considering the lurid history of public-land management in the nineteenth century, is indeed impressive.

In the third place, the leaders of the Forest Service are convinced, on the whole, that the field men are responding to the leadership from the center. Admittedly, the leaders are not unbiased judges. Furthermore, they would feel no sense of disobedience if they made few efforts to exert control. But the leadership of the Washington office is vigorous. And the existence of great rifts between Washington and the field could hardly be completely concealed from an investigator talking informally with many people over long periods and at all levels in the organization. Nor is it likely that the field men have deceived the leaders in Washington into thinking the field people are doing what Washington wants when, in fact, they are doing something quite different; the methods by which such departures would be signaled to the center are too numerous and too sensitive. The consensus among Washington officials of the Forest Service that field commanders down to and including the Rangers are carrying out headquarters policies must therefore be regarded as evidence of organizational integration.

In the fourth place, in the large majority—but not in all, by any means—of the formal appeals from Ranger decisions by forest users, and of inquiries by Members of Congress on behalf of constituents displeased with decisions and actions of forest officers, the Rangers are upheld by their superiors. As noted earlier, this is not simply because of the predisposition of higher levels to back up their subordinates no matter what the subordinates do; such a policy would be too costly in terms of Congressional goodwill and public support, would encourage field men to make decisions without reference to direction from above, and would lose opportunities to train members of the agency in approved patterns of behavior. It may therefore be taken as a sign that the men in the field do adhere quite conscientiously to instructions from their leaders. This consonance between what the top levels say and the lower levels do is here interpreted as indicating the organization for administration of the national forests is highly integrated.

In the fifth place, the frequent transfer of personnel works successfully. A man shifted from one Ranger district to another, from one national forest to another, even from one region to another, is able to take up where his predecessor left off with a minimum of disruption. Were this not the case, the frequent movement of men would have the agency and its clientele in a constant state of confusion. But the unity of the agency is great enough to permit each member to fit into the position vacated by his predecessor quickly and smoothly.

In the sixth place, it is the author's impression after his visits and conversations throughout the country that the techniques of integration have been effective. One of the most striking conclusions about the Forest Service is the degree of similarity among the men in it—their love of outdoor life; their pride in the Forest Service; their habit of taking the long view of things; their patience; their acknowledgment of their obligations to the local users of the national forests; their acceptance of the inevitability of conflict growing out of differences among the many users of the national forests, and between the national interest as against local or special interest; their enjoyment of the variety in Ranger district administration as compared with the narrower scope of industrial forestry; their willingness to do more than is legally required of them in order to get their jobs done. Everywhere, they answer questions about their actions in terms of "policy guidelines." All of them emphasize "multiple-use management" (with apparent unawareness of its obscurity) as the cardinal principle underlying every decision they make about the handling of their districts. Not one failed to mention "fiscal integrity" as the cornerstone of administrative practice. Nor did any get through the interview without at least one reference, directly or by paraphrase, to Gifford Pinchot's creed proclaimed in 1905[1] that the national forests must be managed for the greatest good of the greatest number in the long run (again, with no evident recognition of this slogan's ambiguities). Problems discovered on one district were posed to other Rangers to find out how they would approach them; in general, they suggested very similar solutions, offered similar justifications for their suggestions, and produced instances of a comparable kind from their own experience. Their answers, their questions, their arguments, their explanations, as far as their work is concerned, show common outlines. It is not impossible that this pattern is entirely fortuitous, but it is highly improbable. Far more persuasive is the inference that the modes of integration have had many of the effects they were intended to produce.

By any criterion, then, it must be judged that the leaders of the Forest Service have vanquished the tendencies toward fragmentation.

The Strategies of Conquest

A CONTINUOUS BATTLE

This victory, however, is never won once and for all. The centrifugal tendencies described earlier are not momentary phenomena, which, when counteracted, cease to operate.

[1] See n. 19, p. 84, above.

They are merely held in check by integrative forces greater than they, and every lapse of integrative power is promptly followed by symptoms of dissolution, as was the case, for instance, when the control of Forest Service policy slipped from Washington into the hands of the regional foresters.[2] It takes constant effort to keep the members of a large organization responsive to leadership from the center.

It also takes a high degree of adaptability. The environment of a large organization is not constant; for example, the Forest Service was able to acquire the services of the men with the highest academic records when it was practically the only employer of foresters during the depression of the thirties, but it has had to compete vigorously with private firms paying higher wages during the post-World War II period, to accept men of less distinguished scholastic achievement, and adjust its own training and indoctrination accordingly. Promotion and placement policies, too, have been adapted to hold personnel with new possibilities of lucrative employment on the outside. Internally, the Service has been confronted with the problem of approaching retirement by large numbers of professional men recruited in quantity all at the same time when the Service was expanded rapidly to handle the Civilian Conservation Corps during the New Deal; some younger men will soon be advancing to leadership positions far more quickly than was ever before the case. Locally, communities that welcomed the Forest Service when it was buying property that had become unproductive and almost worthless during the Great Depression now eye longingly national forests made highly valuable by expert management and improving timber markets. In the field, a growing volume of paperwork has been tying district Rangers more and more to their desks, compelling adoption of new practices to free them for more work in the woods.

[2] See n. 17, p. 80.

These ever-changing conditions mean the methods of one decade may not suffice in another. The challenges to unity never disappear; they can only be held in abeyance. The Forest Service has succeeded in suppressing them by attacking each one as it develops. Every tendency toward fragmentation has been met by a strategy to nullify it. Each centrifugal thrust has been counterbalanced.

HIERARCHICAL SPECIALIZATION

The centrifugal effects of intensive functional specialization and problems of communication, for instance—both products of the size and complexity of the Forest Service job, and of the organization required to perform its tasks[3]—have been contravened by what might best be termed "hierarchical specialization." In a general sense, this means (1) concentration by the Washington office on relations between the Forest Service and other organizations and institutions—the Department of Agriculture, the Bureau of the Budget, the Civil Service Commission, the General Services Administration, the General Accounting Office, the Department of the Interior, Congress, and others—and on the framing of broad policy statements of Service-wide effect, generally referred to as "guidelines"; (2) specialization by the regional offices in the adaptation of Service guidelines to regional conditions; and (3) application of the policies to concrete situations by the forest supervisors and district Rangers in the field. But hierarchical specialization means more than this alone. It also describes the position of every line officer as a commander of his own unit. A Ranger's superiors may offer suggestions, apply influence, or even issue orders to him, but, unless he is officially relieved of his command, he issues all instructions to his subordinates and operators. The solidity of his authority is protected against erosion by visi-

[3] Chapters II and III.

tors from above. Those who forget this principle and try to direct activities on a Ranger district without going through the Ranger can be sure of a swift and vehement objection by the field officer. As a matter of fact, a Ranger is expected to defend his position; he would be derelict if he did not. Hierarchical specialization works at all levels.

The existence, at each level, of a single, determinate individual formally empowered to issue decisions with respect to all functions—decisions not subject to further appeal at the same level—means that the competing claims of the several functional specialties will often be judged in terms of more general criteria of decision. It also relieves higher headquarters of torrents of detail from below that would otherwise impede concentration on the full-time task of maintaining integration. At the same time, it safeguards field units against incessant intervention by functional specialists from above, unco-ordinated intervention that could result in hopeless confusion in the field. And it cuts down the problems of communication by establishing, close to the field, "switchboards" in which general directives are adapted to the specific conditions of limited areas, and in which inconsistencies are often discovered and eliminated before instructions take effect upon field personnel. Hierarchy is virtually universal administrative response to the problems of scale in organization; indeed, it is practically a synonym for organization. It creates problems of its own, and some tendencies toward fragmentation derive from it. But it is certainly one of the ways the Forest Service has avoided the splintering effects of other characteristics of a large and complex agency.

MULTIPLICATION OF CONTROLS

Hierarchical specialization also accounts in large part for the pivotal position of the Rangers in national forest administration, and for the ensuing problems of integrating Ranger decisions and actions. These fragmentative forces, in turn, have been answered specifically by two other modes of integration: the elaboration of the body of preformed decisions and the refinement of the methods of detecting and discouraging deviation from them. As the intensification of forest management and the burgeoning of Forest Service activities heightened centrifugal tendencies, the central controls were multiplied.

The extent of the multiplication is indicated by the growth of the *Forest Service Manual.* Originally, forest officers were guided in their administration of the national forests by *The Use Book*, a volume of some 225 pages that included everything from a history of the agency, a description of its major policies and procedures, and the major statutes and judicial decisions concerning the national forests, to instructions on correspondence and filing. Year by year, the volume grew, eventually to evolve into four of the seven volumes of the present *Manual*, much expanded and far more detailed. It was also elaborated by the regional office and supervisors' office supplements, by the technical handbooks and manuals, by the extensive system of work load and financial planning, and by the development of all manner of technical plans, described earlier. What began as a relatively short document a man could slip into his pocket in the field turned into a library within a couple of generations as the Forest Service leadership labored to ensure the districts would be run as components of a unified agency, and not as autonomous microcosms of the Forest Service.

Such an outpouring of materials can impede attainment of conformity instead of encouraging it if the volume of materials and the complications of posting changes impose such a burden on field officers that these men give up all hope of keeping abreast of them. In the Forest Service, the revision of the *Manual* now in progress represents an effort to forestall the growing possibility that the sheer quantity of regulations will defeat the purpose for which they are issued. In addition, Rangers have customarily been sympathetically received when they consult about individual cases with specialists at higher levels who can be depended upon to know the current status of the rules applicable to their specialties. Furthermore, with the multiplication of preformed decisions, the devices for detecting deviation were likewise refined and intensified to fit them to the growing body of rules and regulations and other communications flowing from the center; most of them had been in use in the Forest Service from its very inception, but they were brought to a level of accuracy and sensitivity previously unknown.

It was not simply, or even primarily, to reduce field discretion that communications to the field were multiplied and elaborated, and backed up by an increasingly searching system of keeping close check on compliance with them. Many other considerations contributed to these practices. More intensive utilization of national forest resources and consequent functional specialization of staff, enactment of additional statutes, amendment of existing laws, changes in regulations necessitated by altered circumstances and by lessons learned by experience, and modifications of policy dictated by changes in the political complexion of the government (particularly of the Secretary of Agriculture) all evoked new instructions and interpretations. Increases in technical knowledge about all phases of forestry resulted in expansions of existing manuals and handbooks, and in additions to the list and to professional libraries. The rise of scientific management gave rise to meticulous work measurement systems, given added impetus in the depression of the 'Thirties, when public agencies had to be able to defend their requests for appropriations with all the data they could muster in order to win funds from Congresses trying desperately to cut the size of the federal budget. In order to be certain the new measures and processes were functioning properly, and to feed back the detailed information on which the successful operation of many of them depends, the methods of detection had to be improved and stepped up. Hence the multiplication of controls. Nonetheless, although the sources and motivations of these developments were certainly plural, the specific effects of the developments were to offset, by contracting the discretion of field officers, the impulses toward fragmentation stemming from the pivotal role of the district Rangers.[4]

ELIMINATION OF BARRIERS TO COMMUNICATION

Increases in the number and specificity of preformed decisions at all levels above the Ranger district diminish one of the problems of internal communication in the administration of the national forests: the ambiguities of instructions framed in general terms and the opportunities this gives men in the field to construe and apply the instructions in widely varied ways. As noted earlier, there is no way to eliminate this latitude, but it has certainly been narrowed in the course of the years.

A different set of factors negates the centrifugal tendencies of the other communications problems of the Forest Service: inconsistent directives, and obstacles introduced by distance (both geographical and sociological). As far as geography is concerned, the Forest Service has been quick to test carefully and install promptly every device available to speed travel and communication; even the remotest stations have been brought into frequent contact with their next higher headquarters by the alacrity with which the Service puts latest technological developments to use. As far as the other difficulties are concerned, more subtle measures proved necessary.

[4] See pp. 47-64.

All the other difficulties spring from one characteristic of the organization, namely, the extent of specialization of personnel—by hierarchical level, by the functions they perform, by the areas they administer. This is the origin of the inconsistent instructions, and of the status, linguistic, and attitudinal barriers discussed in an earlier chapter. What tends to minimize these differences may be termed the "homogenization" of professional personnel by the Forest Service.

One of the principal means to this end is the use of professional foresters in all types of specialized jobs. Not only are line officers, timber management staff men, and fire control specialists all foresters, as one would expect, but so are the specialists in range management (who could be trained in animal husbandry as logically as in forestry), wildlife management (who could be biologists), personnel management, administrative management (whose academic training could well be in public administration or industrial engineering), information and education, budgeting, recreation management (all of whom could be trained in their particular fields), and in other functions. Probably 90 per cent of the professional men engaged in the administration of the national forests are foresters.[5] Since difficulties of communication among specialists stem from the perspectives of their jobs, as well as of their professions, establishing the predominance of a single profession throughout the organization

[5] From the Washington office came the following comment:
"The paragraph, as written, does not mention the fact that the Forest Service does employ, on a recurrent planned basis, through Civil Service procedures, men in the following fields: (1) Engineers—several disciplines, including civil, mechanical and electrical. (2) Business Management and Accounting graduates. These men have approved career ladders and are initially assigned to national forests or ranger districts. (3) Specialists in any of the engineering or biological sciences may be employed from time to time when their services are needed to carry out either administrative or research programs. This employment does not change your basic point, but our employment is a little broader than you imply."

does not end the troubles entirely. But it certainly *reduces* them materially.

In-service indoctrination and training further standardize universes of discourse, attitudes, skills, and interpretations. In particular, the emphasis on multiple-use management of natural resources, and on the philosophy as well as the practice of multiple-use, broaden the appreciation of the context in which every Forest Service job is performed. The "tunnel vision" that normally afflicts experts in one phase of a program is corrected. While the men do not emerge from their training in utter uniformity, they inevitably carry away with them an understanding of many aspects of the Forest Service mission, which leads them to consult with their fellow-specialists, and to take the others' needs and views into consideration when framing their own projects.

Rotation of professional personnel by lateral transfer and by "diagonal" promotion—i.e., by moving men to different areas and functions as well as to different levels—and by assignment of men to tours of staff duty at administrative levels above the levels at which they will serve in line capacities, add to the "homogenization" of personnel. Rotation of this kind, it should be noted, presupposes a substantial measure of uniformity; otherwise, officers would not be interchangeable, and could not promptly and efficiently assume the duties of those they replace. Frequent movement of personnel thus depends on the common professional background and training of Forest Service personnel. At the same time, it amplifies the effects of these traits in annulling the divisive tendencies of specialization.

For it permits the members of the agency to get to know each other, and thus to develop a mutual frame of reference. It familiarizes them with the range of Forest Service functions. It acquaints them with the manifold territorial conditions under which national forest administration must be carried on. It exposes them to the perspectives of their superiors before they move into line positions. And, for all these reasons, it breaks down the insularity of specialization and the barriers between administrative levels.

Field inspections also foster common viewpoints and attitudes. For one thing, they bring officers from many levels into personal contact with each other, often under quite informal circumstances; rank differences all but disappear when there are camping trips, or after-hours social visits, and official relationships give way to greater spontaneity. Secondly, the stress in inspections is on training, and the inspectors may be said to constitute an itinerant school that transports a basic curriculum taught by the same staff to field units throughout the country. An inspection tour is a broadening experience for the men who are inspected (and for the inspectors as well). The process contributes to breaking down the parochialism of everyone in the Forest Service.

Besides "homogenizing" personnel, the Forest Service has instituted formal procedures to prevent the issuance of conflicting decisions to be followed by field officers. At every level above the Ranger districts, lateral review and clearance of all issuances impinging on more than one activity are standard, and many plans and instructions must collect the signatures of all the specialists involved. Thus, differences of opinion are ironed out prior to promulgation, safeguarding field men from the confusion of irreconcilable communications from headquarters, and safeguarding the organization from the fragmentative consequences of having each Ranger resolve the inconsistencies in his own way.

Few of the foregoing practices were adopted with problems of communication in mind; they serve other purposes, discussed in previous chapters. Yet they do overcome very specifically the tendencies toward fragmentation engendered by barriers to communication. The barriers are by no means down altogether, but they are lower everywhere and completely breached in many places as a result.

NEUTRALIZING THE FORCES OF LOCALISM

Rotation of personnel is a versatile administrative tool that works against several centrifugal tendencies. It was just observed that it helps pierce the communication barriers tending to isolate members of the Forest Service from each other. In addition, it is a specific antidote to the capture of Rangers by the forces of localism, both those inside the Forest Service (the behavioral norms of face-to-face work groups) and external to it ("capture" by local populations).

It works against the appearance of fixed group norms in conflict with policies and procedures of the Service in two ways. One is to discourage their appearance: the frequent change of field commanders maintains a state of moderate flux and uncertainty in the permanent sub-Ranger and labor forces on Ranger districts. Each new Ranger comes with his own set of preferences, habits, and expectations, continually disrupting or working changes in the customs of the local employees. The work norms are thus kept sufficiently fluid to prevent the Rangers from becoming their helpless prisoners.

The other way is to inhibit complete identification of the Rangers with their aides and other subordinates. The Rangers move around enough to avoid deep involvement in, or commitment to, any one local set of norms. Because of their broad experience, they tend to follow agency rules rather closely; the Service-wide methods of operation furnish them with guidelines applicable everywhere. Every individual placed in a strange work situation generally falls back on his background and training, and applies them to the new conditions. The background and training of the Rangers are not local, and they are often shifted before they fully internalize prevailing local work folkways.

Moreover, the permanent local work forces, being exposed to so many commanders over the years, come to know a great deal about the Forest Service way of doing things. Some of the more experienced hands thus come to keep the Rangers on the proper path instead of leading them in deviant directions.

As a result, the patterns of "informal organization" in national forest administration are rarely at odds with policies enunciated by the higher levels.

The practice of transferring men rapidly, particularly in the early stages of their careers, also counterbalances the danger of their being "captured" by the communities in which they live and work. To be sure, the Forest Service encourages its men to recognize and understand the concerns of their communities, and to take part in community affairs. But there have been instances in which the utility of a good forest officer has been gravely reduced because he became so enmeshed in local affairs that he could not properly discharge his responsibilities as a representative of the agency and the agency's view of the public interest. Hence, foresters are rotated fairly often to keep them from absorbing the premises of decision of their localities to the exclusion of the Service's premises, from becoming purely spokesmen for local special interests. In a previous chapter, it was noted that the professional and organizational atmosphere of the Service becomes the most stable element many field men know in their early years, and their bonds with the organization are highly developed and deeply ingrained by the time they settle down in a single locale, and in a single job, for any extended interval. In this way, the risks of "capture" are substantially lessened; loyalities to the agency are so well nurtured that they offset competing loyalties to other groups and symbols.[6]

[6] Not invariably, however, the pressures on a field officer and his family can become quite intense, and any vacillation or hesitancy at higher levels in relations with special interests may be interpreted as signals for compromise below. This is alleged to be particularly true of range management in the West, where the livestock industry is highly

As is true of most Forest Service practices, rotation was instituted for many reasons besides those cited. Nevertheless, the success of the Service leadership in suppressing the thrusts toward disintegration must be ascribed in part to the effects of the practice whether the effects were intended or not.

MANIPULATING PERSONAL PREFERENCES AND PERSPECTIVES

The influences on the decisions and behavior of Rangers, it has been seen, come from four sources: higher headquarters, face-to-face work groups, local interests, and from "inside" the Rangers themselves. The goal of the leaders of the Forest Service is to make the influences from higher headquarters internally consistent, and then to have them prevail over all those from the other sources when the others conflict with their own. The strategies analyzed up to this point operate chiefly by strengthening the signals to the Rangers from higher headquarters, and by neutralizing (i.e., blocking or overwhelming) those from face-to-face work

influential and still displays attitudes stemming from virtually uncontrolled use of the public ranges for many years prior to the establishment of the Forest Service and the Grazing Service. Indeed, as a matter of national policy, range management programs are tied closely to local customs, as regards planning and assessment of the carrying capacity of land, through extensive consultation with local associations of grazers. Moreover, some young foresters depend heavily on experienced range riders hired by the local associations. Under these conditions, local standards play a large part in official behavior.

It is therefore not surprising that a couple of the Rangers interviewed for this study indicated they would have preferred smaller herds on their districts, but thought reductions would for tactical reasons have to be achieved gradually. It seems highly probable, however, that the Rangers *would* exert all their energies to carry out "tough" policies if they were *directed* to do so, even if this subjected them to extreme pressure in their communities. If it is true that the forces of localism remain strong in some areas and for some functions, it is not because the Rangers have surrendered to them. The administrative practices of the Forest Service have prepared them to do what is required of them under any circumstances. When the Rangers reach accommodations with local interests on major issues, it is more often because this seems to them to be the agency strategy than because they have been captured.

groups and local interests. But there is still the possibility that the personal preferences of the Rangers with respect to the work they have to do may run contrary to the desires of the organization and win out over the organization-manipulated influences. A battery of influences, referred to in an earlier section of this volume as modes of implanting the will and the capacity to conform to preformed decisions, prevent this possibility from being realized very often.

It operates partly by overriding the personal preferences of the Rangers—by inducing them to conform to communications transmitted through official channels no matter what they feel about the actions thus required of them. This is often accomplished by arousing anxiety or hope. All people are presumably desirous of retaining their jobs, getting pay increases and promotions, and enjoying the other material and intangible benefits of organizational life; they are therefore vulnerable to threats of being deprived of some or all of these, and responsive to the promises of reward. Furthermore, many people seem to grow anxious when they find themselves at odds with the weight of opinions they respect; it is more reassuring to be with the authorities and with the group than all alone against them.[7] The officially promulgated policies of an organization like the Forest Service generally carry the prestige of its leaders and the support of most of its members. The dissenter, exposed, is apt to feel uneasy, embarrassed; the conformist, secure, reinforced, accepted. So anxiety and hope lead men to adhere to instructions and established procedures though they have doubts or even strong negative feelings about them.

Personal preferences are overcome also by arousing shame and guilt in those who submit to such preferences in contra-

[7] See S. E. Asch, "Effects of Group Pressure upon the Modification and Distortion of Judgments," in D. Cartwright and A. Zander, *Group Dynamics: Research and Theory* (White Plains: Row, Peterson & Co., 1953).

vention of the official position and practice of their agency. Unlike the sources of hope for reward and of anxiety about punishment just mentioned, which are exploited but not created by the organization, sources of guilt and shame about deviating from preformed decisions must be instilled in organization members; otherwise, personnel would feel no more concern about undetected deviations than, say, about jaywalking. In the Forest Service, the cultivation of identification with the agency as a whole develops a foundation for both guilt and shame. As the values of the Service are internalized by each individual, the importance of obedience to the survival of the organization becomes paramount. As the habits of the hierarchy are absorbed, the "right" of superiors to ask, expect, and receive obedience is also assimilated. As personal ties with professional colleagues become more intimate, the desire not to let one's friends down grows stronger.[8] Once these sentiments appear, men who violate the standards and principles of the agency begin to feel twinges of conscience and a little ashamed; from that point on, they keep themselves in line, even when compliance with preformed agency decisions collides with their personal inclinations.

Overriding personal inclinations, whether by playing on hopes and anxieties, or by stimulating guilt and shame, does not *eliminate* them as challenges to the unity of the organization for national forest administration; it simply opposes them with a stronger force. The preferences remain. The

[8] See, e. g., E. A. Shils and M. Janowitz, "Cohesion and Disintegration in the Wehrmacht," *Public Opinion Quarterly*, Vol. 12, No. 2 (Summer 1948), pp. 280-315: "It is the main hypothesis of this paper . . . that the unity of the German Army was in fact sustained only to a very slight extent by the National Socialist political convictions of its members, and that more important in the motivation of the determined resistance of the German soldier was the steady satisfaction of certain *primary* personality demands afforded by the social organization of the army," (p. 281), See also Stouffer, *et al.*, *The American Soldier* (Princeton: Princeton University Press, 1952), Vol. I, pp. 410-420.

administrative strategies of the Forest Service do not stop with this; they also turn the preferences themselves to harmonize with the objectives of the leaders. Instead of just overpowering troublesome preferences, the Forest Service nourishes and harnesses useful ones. That is, the techniques of integration earlier identified as developing the will and the capacity to conform are positive as well as negative in effect. They do more than elicit reluctant obedience. They do more than persuade each Ranger to assign higher priorities to the wishes of the organization than to his own. They actually infuse into the forest officers the desired patterns of action in the management of their districts, so that the Rangers handle most situations precisely as their superiors would direct them to if their superiors stood looking over their shoulders, supervising every detail. To overstate the case, their decisions are predetermined. From the Rangers' point of view, they are not obeying orders or responding to cues when they take action on their districts; they are exercising their own initiative. It is not compulsion or inducement or persuasion that moves them; it is their own wills. Speaking figuratively, it would not occur to them that there is any other proper way to run their areas. That is why the Rangers generally feel so free and independent, see themselves as "their own bosses," characterize themselves as "kings in their own domains." After all, they are doing just what they want in the light of their perceptions.

The preferences and perspectives of the Rangers are certainly not established by Forest Service manipulations alone. Forest Service control reaches only a segment of them. The men in the agency are a long way from being automata. But these influences are strong enough to nullify or supplant many of the tendencies toward excessive diversity arising from individual differences in the premises of decision.

The consequences of this strategy are far-reaching. In the last analysis, all influences on administrative behavior are filtered through a screen of individual values, concepts, and images. Some signals are screened out, some come through in full force, some are modified or attenuated. To the extent the leaders of an organization can manipulate the screen, they can increase the receptivity of field personnel to organization directives, decrease their receptivity to outside influences. It is an answer not just to one centrifugal tendency, but to many.

EXPLOITATION OF FORTUITOUS FACTORS

Clearly, the efforts of Forest Service leaders to influence the administrative behavior of field personnel so as to keep the agency an integrated, a unified, organization is a battle against conditions over which the Forest Service itself has little control: the course of history, the dispersion of the properties entrusted to it, the nature of large-scale organizations, the strengths and weaknesses of individual men, competition with private industry for recruitment of foresters, and many others. Against this array of challenges, it has made a planned, deliberate campaign, using every instrument at its disposal.

However, there are fortuitous factors working to its advantage as well as against its unity. For instance, some of the leaders believe that the assaults on Forest Service policy by lumbermen anxious to cut more heavily in the national forests, and by stockmen seeking to graze more animals than the Service thinks the ranges on the national forests can safely sustain, have actually heightened the enthusiasm and morale of many of the men and strengthened their bonds with the agency. For these demands have given forest officers a sense of engagement in a crusade on behalf of the public interest. Their duties are elevated from routine forest management to safeguarding the economic, and perhaps even the military, security of the nation. They know the excitement of a contest, the risks of the firing line, the immense gratifications of performing an important service for the whole people. They are placed squarely in the tradition of Gifford Pinchot, who battled furiously to prevent what he considered the depredation of public property.

Propaganda hurled against the Forest Service doubtless makes some field officers uneasy, and leads them to question the policies they are required to carry out, and perhaps to execute those policies less vigorously than they otherwise would; for this reason, the agency answers the attacks with information and education. At the same time, the attacks endow its work with a drama it might otherwise lack, a sense of urgency that makes its members feel they must stand together. Statesmen have known for thousands of years that an external menace can unite the people of a country as no other single event can; on a small scale, the threat to the Forest Service has had a similar unifying effect.

Another kind of unplanned assistance comes from the very nature of the resources with which the Forest Service deals. Biological processes can be encouraged, assisted, and favored, but they cannot be significantly hurried. Under the most favorable conditions, it takes 20 to 25 years for a tree to reach pulpwood size, it may take twice that long in regions where the growing season is short or the soils poor; and it generally takes three to five times as long to bring a saw timber species to maturity. Similarly, it takes decades of careful management to bring an abused range back to a healthy state. Small wonder, then, that foresters develop the habit of

> thinking in terms of decades and centuries, rather than of days and years, working on with persistent patience. The long look ahead is perhaps the most distinctive characteristic of our profession. Often the laymen cannot sympathize with the forester here, or cannot comprehend this mental approach. Yet with us it is a *sine qua non.*[9]

[9] W. Mulford, *The Professional Attitude of Foresters* (Duke University: School of Forestry Lectures, No. 2, May, 1941), p. 13.

> Forestry is forced by the very nature of the crop with which it deals to take the long view. Every time a forester or timber-land owner plants a tree, conducts a logging operation, or prepares a management plan he is influencing the course of events for the next 10, 50, or 100 years.[10]

In such a frame of reference, decisions and actions that threaten to generate friction between headquarters and the field can be deferred—a delay of a year, or two, or five, in many instances, is not likely to be critical. If a Ranger, sensitive to local sentiment, fears that a tree-spraying program, or road construction or logging on a municipal watershed, or a curtailment of grazing or other permits will stimulate local opposition, it does no great harm for him to put them off for a while, until conditions become (or can be made) more favorable. These tolerances avert many potential tensions between headquarters and field officers. They afford Rangers opportunities to reconcile or resolve some of the competing demands on them, and they permit Ranger errors to be effectively corrected even if they are not caught at once.

A third kind of fortuitous aid is technological innovation, from which the Forest Service has benefited greatly. The revolution in transportation and communications has made possible increased surveillance of field units. Business machines permit centralized record-keeping. Automotive equipment makes possible accurate checks on travel distances and times. Moreover, more intensive management of forest resources has been facilitated by tractors, power pumps, radio, and the airplane (for "smokejumpers," for fire detection, for aerial spraying). It is to the credit of the Forest Service that serviceable new inventions are promptly utilized. Nevertheless, the Service itself is not responsible for their invention.

[10] Editorial, *Journal of Forestry*, Vol. 42, No. 12 (December, 1944), p. 859.

Fourth, the political situation and historical position of the Forest Service minimize a variety of forces that might render it far more vulnerable to the pressures of special interests than it now is. The demands of the lumber industry, for example, and grazers, and other forest users, have been balanced by the outcries of a strong, well-established, and devoted conservation movement. The location of the Service in the Department of Agriculture rather than the Department of Interior has shielded it from the influences of some of its critics, whose lines of access to the latter department, which has jurisdiction over the public domain (and ran the forest reserves until the creation of the Forest Service in Agriculture in 1905), are extremely well developed; these critics are less influential in the Department of Agriculture, which has a rather different constituency.

Similarly, the main Congressional lines of access of the principal critics—lumber and grazing interests and other public-land interest groups—do not run to the agricultural committees to which the Forest Service reports, or to the agricultural appropriations subcommittees to which it reported until recently; the membership of these bodies is weighted toward the farm states rather than the lumber and grazing states and the states with massive blocs of federal lands, so that the Forest Service has been somewhat less susceptible to its chief critics' pressures than it might otherwise have been. Partly for this reason, the statutes under which it operates are generous in their grants of discretion and powers. Forest Service strategists have been quick to recognize the benefits of the agency's position and relationships, and work hard to preserve them as they are, but today's leaders are the beneficiaries of past events rather than designers of these advantages.[11] Conceivably, if the agency

[11] Actually, it is no accident that the Forest Service enjoys the strategic advantages of location that it now does; Gifford Pinchot clearly planned to establish the new bureau in an environment free from the

were not so fortunately situated, it might be compelled to let its field units arrive at their own policy accommodations with the interests in their localities so as to avoid provoking storms that could destroy it. Protected as it is, the Service can afford to build the field receptivity to central leadership that is such a striking feature of its internal operations.[12]

Where, in short, the environment of the Forest Service works against unity, it is neutralized. Where the environment may be actively employed to promote unity, it is exploited. Where the environment helps accidentally, it is enjoyed.

THE INTERACTION OF INFLUENCES

This catalogue of the ways in which each of the tendencies toward fragmentation in the Forest Service is negated is illustrative rather than exhaustive. However, it covers the major strategies and their consequences, and demonstrates the methods by which the Forest Service has remained a well-integrated, highly unified organization in spite of all the forces that might be expected to set every Ranger running his district in a fashion unrelated to the operations of his colleagues or the wishes of officials in Washington.

On the other hand, it is extremely difficult to determine which of the modes of integration is prior to the others. Their

forces that had been at work within and upon the land management agencies of the Department of the Interior. As early as 1903, President Theodore Roosevelt established a committee to recommend to him improvements in the organization of "executive government work"; Pinchot was secretary of the committee, and it is therefore hardly surprising that it urged the transfer of forest reserves, and of all related activities, from Interior to Agriculture. See G. Pinchot, *Breaking New Ground* (New York: Harcourt, Brace & Co., 1947), p. 242. Pinchot obviously saw the benefits to the agency of being in the one department rather than the other.

[12] I am indebted to Professors Wallace S. Sayre, of Columbia University, and James W. Fesler, of Yale University, for pointing this factor out to me.

effects are so intermingled that no variable seems clearly independent.

Logically, manipulation of personal preferences and perspectives appears to be the fundamental mechanism. When the forces "inside" a man drive him to act in the organizationally approved manner, there is little need to issue preformed decisions and to check his performance for deviation; everything he does can be predicted. Extraneous local considerations, and other "pulls" that might produce deviant behavior, would be rejected by the individual. On his own, without prompting and without compulsion, he would do what was wanted of him.

Fortunately, men are not quite so malleable. They cannot be programmed like electronic computers. In an imaginary *Brave New World*,[13] their minds might be molded entirely by high officials until they are as alike as automobiles rolling off an assembly line. In the real world, despite the awesome modern methods of controlling the mind, men are still infinitely varied, the products of their own unique heritages and experiences. Clearly, the leaders of all organizations must continue to rely on influences exerted "upon" them to control their behavior.

Moreover, it appears that the influences "on" Rangers are major factors in achieving whatever manipulation of personal preferences and perspectives is possible. The Rangers want to do the very things the Forest Service wants them to do, and are able to do them, because these are the decisions and actions that become second nature to them as a result of years of obedience. Recruits may conform to official requirements against their instincts, their desires, or even their judgment at the outset, as the techniques of overriding their personal feelings are applied. After a while, though,

[13] A. Huxley, *Brave New World* (New York: The Modern Library, 1956).

they come to appreciate the reasons for the policies they execute, and they come to comply with instructions habitually, instinctively, naturally. If there were no preformed decisions and devices for discovering and correcting deviations from them, it seems likely many of the patterns of behavior that become habitual would never be established; that is, "internal" forces may well depend, at least in part, on external influences.

At the same time, the operations of the external influences may depend on successful manipulation of the personal preferences and perspectives "inside" the Rangers. Systematic selection and training of personnel, and procedures for building identification with the Forest Service, increase Ranger receptivity to the communications of the central office. They broaden what Simon [14] has called the "zone of acceptance," inculcating in field officers the predisposition to respond primarily to cues and signals from the leaders of the agency, and to resist conflicting influences from other sources. Without this process, there would be none of the obedience through which prescribed patterns of values and action are "internalized."

The influences "upon" and the impulses "inside" the members of organizations thus seem inseparable. The Rangers internalize Service modes of behavior because they obey orders, yet they accept many orders because their preferences and perspectives have been manipulated.

Empirical tests might suggest whether one or the other is prior, the independent as against the dependent variable. For example, a number of Rangers could be designated an experimental group and exempted from some of the formal influences to which they are now subject. Funds might then be allocated to them on a lump-sum basis instead of in

[14] H. A. Simon, *Administrative Behavior* (New York: The Macmillan Co., 1947, 1957).

itemized detail. Rather than a small library, only a few essential policy guidelines might be given them as instructions and standards. The number of reports could be reduced. Diaries might be abandoned. Inspections could be made less frequently. A panel of judges consisting of high officers of the Forest Service could periodically investigate and compare performance on the experimental districts with performance on others chosen as a control group. If the records then showed the experimental group deviating markedly from announced policy, this would indicate the indispensability of the influences "on" the Rangers; if the records showed no significant variation between the experimental group and the control group, it would suggest that some of those influences are perhaps superfluous. It would be inappropriate to attempt to set forth here a design covering every phase of such an experiment, but there is every reason to believe such an experiment is feasible and likely to be fruitful. Statutory restrictions and requirements and the burden of day-to-day operations may render the Forest Service an unsuitable laboratory for such a test, in which case another organization not similarly encumbered would have to be found. Until empirical data are gathered, however, the question of the relative importance of the major categories of influences on Ranger behavior remains open, and the contradiction of any opinion is as plausible as the opinion itself.

This ambiguity beclouds one of the great issues of administration: the nature of centralization and decentralization. If experimentation discloses that field behavior can be controlled as effectively by inculcating the fact and value premises of central headquarters upon the minds of field men *without* extensive use of close supervisory and enforcement procedures, as is possible *with* these devices, then an organization which gives every indication of decentralization by all the usual indices [15] may in fact be as fully governed from the center as one without these visible paraphernalia of central direction. And, conversely, the elaboration of formal controls may indicate only that field men in the organization in question are not responsive to central leadership, and that previous attempts by the leadership to influence their behavior have failed, necessitating still more and tighter efforts at direction; thus, the seemingly centralized organization may, from a behavioral standpoint, be more decentralized than one lacking the traditional procedural manifestations of centralization. The usual criteria stress external forms and tend to neglect actual behavior. If actual behavior is shaped more by factors "inside" individuals than by external forms, and if the internal factors are produced by something other than formal directives and controls, then the traditional criteria may often be misleading. The uncertainties muddy the analysis of the problem.

No matter what analytical doubts are raised by the interaction of influences, however, two things are quite clear. One is that the centrifugal tendencies confronting the leadership of the Forest Service have been vanquished. The other is that the conscious and deliberate strategies of the leaders, though aided by fortuitous factors, are largely responsible for the victory. This much, at least, is unequivocal.

[15] These are gathered from the literature and summarized in D. B. Truman, *Administrative Decentralization* (Chicago: University of Chicago Press, 1940), pp. 56-59, as follows: (1) the extent to which "local coordinators" supervise field specialists; (2) "the frequency with which field offices refer matters to headquarters for decision;" (3) "the number and specificity of general regulations or special directions under which the field agents work;" (4) "the provision for appeal from the decision of field agents;" and (5) the number of decisions on individual cases made by field men, and the variety of duties performed by field men. See also J. W. Fesler, "Field Organization," in F. M. Marx (ed.), *Elements of Public Administration* (New York: Prentice-Hall, 1946), pp. 264-93.

The Hazards of Managerial Success

A PROBLEM OF ETHICS

The realization that successful organizations manipulate the intellects and the wills of their members, as Forest Service experience illustrates, and as has doubtless been true for centuries, has recently produced a flurry of alarm about the ethical implications of such practices. To some observers, this seems to be a threat to the freedom and the dignity of man. At first, the specters conjured up by Huxley [16] and Orwell [17] were considered nothing more than the figments of extraordinarily vivid imaginations. But the growth of propaganda and psychological warfare made them look less fantastic. "Brainwashing" and "subliminal" advertising brought them uncomfortably close to actuality. The awareness that these techniques have counterparts in many familiar and esteemed institutions dawned suddenly and frighteningly on some observers (although the techniques are far older than the modern awareness of them). The protests of Whyte [18] against "the organization man" fell on sympathetic ears.

These fears are not utterly without foundation, but they are based on a partial view of reality. In the first place, as long as men belong to many independent organizations, and as long as they circulate freely among many groups, the probability is slight that any one of these groups—even the one that absorbs most of their time, the one they work for—can reduce them to mere extensions of the wills and purposes of its leaders. (This is undoubtedly why the framers of the Bill of Rights deliberately included freedom of association

[16] Huxley, *op. cit.*

[17] G. Orwell, *Nineteen Eighty-Four* (New York: Harcourt, Brace, 1949).

[18] W. H. Whyte, Jr., *The Organization Man* (New York: Simon and Schuster, 1956).

in their list of fundamental rights.) In the second place, that which is called control of the mind is, when viewed from another standpoint, also termed morality.[19] Conscience, principles, patriotism, honor, devotion to duty and to one's comrades, unswerving justice, compassion, resistance to temptation, refusal to submit to attempted intimidation, self-control, and many other much-admired qualities, are evidences of values, attitudes, and beliefs so deeply ingrained that self-interest, personal desires, and all manner of other stimuli and cues are rendered nugatory as influences on behavior. The same applies to the zeal, conscientiousness, and integrity of the men in the Forest Service; these traits are so thoroughly infused into them that the Service has never been touched by so much as a breath of scandal, although it is the custodian of properties worth hundreds of millions of dollars, has handled many hundreds of millions in receipts and expenditures, and is responsible for a program that was beset by fraud and dishonesty for much of the last third of the nineteenth century. By way of contrast, the failure to instill appropriate habits of thought and action in the minds of American troops in the Korean War has been advanced by some authorities as the explanation of the breakdown of military organization among American prisoners of war, their widespread collaboration with the enemy, and the internecine relationships among them in prisoner-of-war camps; indeed, this record gave rise to a military code of conduct, and to methodical programs to indoctrinate American troops with it.[20]

[19] I am most grateful to Professor Arthur W. Macmahon, of Columbia University, for calling this fact to my attention.

[20] See *The New York Times*, January 6, 1958: "The Russians call it political indoctrination. We call it troop indoctrination.

"However brash it may sound, the United States armed services, especially the Army, feel compelled to teach Americans that in wartime their country comes before self.

"Further, the services feel it is imperative to convince men that their

Administrative success apparently depends on securing a fairly substantial degree of field adherence to the wishes of central headquarters. Securing this conformity, in turn, necessitates a measure of manipulation of mental processes. Undoubtedly, it can be carried to a point menacing to human dignity. But, from all indications, it cannot be eliminated from any organized human activity. It is an intrinsic attribute of large-scale organization.

A PROBLEM OF FLEXIBILITY

A high degree of conformity presents other risks, too. In a dynamic world, policies and procedures must change if an organization is to survive and prosper. Manipulation of mental processes may stifle flexibility. Flexibility depends on the conception of new ideas and the adoption of the best ones. An organization consisting of men who have internalized organizational perspectives, values, and premises may

way of life as free Americans is superior to the tyranny of communism and hence worth fighting for.

"The services, in brief, no longer assume that every man comes to them a diehard patriot.

"Thus we have the strange spectacle of Americans using every weapon in the arsenal of propaganda and psychological warfare—lectures, movies, posters, pamphlets, discussion groups and books—on other Americans.

"This has been going on quietly for the last two years in this country and wherever American troops are stationed from Germany to Japan.

". . . The object is to produce a reliable fighting man."

Recognition of the need for such indoctrination was generated by the fact that "Roughly one out of every three American prisoners collaborated with the Communists in some way, either as informers or as propagandists." None managed to escape because their comrades gave them away. Many were abandoned by their fellows and died of malnutrition. "Discipline among Americans was almost nonexistent. It was a case of dog eat dog for food, cigarettes, blankets, clothes. Many officers and noncommissioned officers refused to accept the responsibility of leadership."

In the Forest Service, analogous collapses of individual integrity and organizational unity have for many years been successfully prevented by techniques comparable to those recently adopted by the armed forces.

well become infertile on the one hand and unreceptive on the other.

Creativity—discovery and invention—is often the fruit of random differences engendered by deliberate experimentation or by spontaneous or accidental exploration. An individual imbued with the spirit of an organization, indoctrinated with its values, committed to its established goals and customary ways, and dedicated to its traditions, is not likely to experiment a great deal, nor even to see the possibilities suggested by unplanned developments.

Receptivity to new ideas in an organization depends on the open-mindedness of its leaders, and on the imaginativeness of all administrative echelons. If the intermediate ranks lack tolerance, novel proposals may be filtered out of the channels of communication long before they reach the top, which deprives the leaders of choice among alternatives; they become trapped by the system. Furthermore, if ideas are transmitted to the leaders, either from inside the organization or from the external environment, and the leaders themselves are products of the organization, having come up through the ranks and absorbed and assimilated the organizational culture, it would not be surprising if they rejected unusual and unorthodox suggestions.[21]

Thus, an organization can be afflicted with a paralyzing rigidity, a stubborn clinging to tried and true methods, familiar goals, established programs. If conditions were stable,

[21] Novelists and others have often observed that the founders of organizations are goal-oriented, while their followers—who preserve the organizations launched by the founders—tend to become organization-centered and rule-oriented, particularly if their organizations flourish. Already, there are signs that this is happening in the Forest Service, that the rough-and-ready ways of the old pioneers have no place in the modern environment, that the skills and strategies of the systematizer are replacing the arts and the fervor of the innovator. A new generation of foresters is appearing, some of whom admit that Gifford Pinchot would probably not get very far if he entered the Service today; he would be too upsetting to the established order.

this policy would be perfectly adequate. But conditions change. Organizations, to survive, must change with them.

The Forest Service, despite its success in injecting its own outlooks into its men, has avoided many of the hazards of success; it has preserved a good deal of its own flexibility. Alert to the dangers, the leaders frequently engage outside consultants to check their methods and procedures, grant leaves for their members to get additional university training (often in fields other than technical forestry), encourage personnel to become active in professional and business and civic associations and societies, open their doors to social science researchers, circulate current literature throughout the agency, and otherwise invite and seek out developments and innovations in all fields; these are calculated efforts to assure a steady influx of new ideas. In addition, field men are periodically canvassed for opinions, suggestions, criticisms and complaints; an atmosphere of informality and candor in relationships between high-ranking officers and field personnel is cultivated; liberal use is made of the government suggestion system for rewarding public employees whose recommendations lead to improved service and/or reduced costs; and men are brought into higher headquarters, including the Washington office, from the field for temporary tours of duty; these are all planned means of fostering transmission of new ideas upward to the very apex of the hierarchy. Finally, the elaboration of staff units represents a decision to promote studies and reflection by men free of line responsibilities and duties, men who are expected to generate new practices, and to prod line administrators into applying them.

The only device the Forest Service has steadfastly declined to employ, at least in national forest administration, is "lateral entry" of personnel—that is, the bringing of men directly into high positions from outside the agency. It has stood firmly by its policy of promotion from within, contending that abandonment of this practice would cost it dearly in morale difficulties, and in administration by men inexperienced in Forest Service problems, while yielding little more in the way of innovation than is now introduced by existing techniques of spurring freshness of viewpoint. Moreover, it is argued, when the Service promotes a man, it knows exactly what his capabilities are; when it brings in an outsider, it deals in unknowns. Lateral entry remains off the extensive list of Forest Service methods of providing stimulation and encouraging receptivity to departures from precedent.

Besides the agency's deliberate steps toward continued flexibility, there are many forms of stimulation it cannot avoid even if it should want to do so. The General Accounting Office may offer suggestions—and criticisms—as part of its auditing function. The Bureau of the Budget makes proposals on the basis of its budget analyses and studies of administrative management generally. The Civil Service Commission, and other overhead agencies of the government, and the staff officers of the Department of Agriculture, are likewise sources of outside advice. The forestry profession generally, and the schools of forestry, advance recommendations and proffer evaluations of performance. Other public land management agencies—the Bureau of Land Management, the Corps of Engineers, the Forestry Division of the Tennessee Valley Authority, bureaus of the Department of Agriculture, state forestry departments and fish and game commissions—present examples, competition, co-operation. Private interests, such as woods-using industries stockmen, hunting and fishing clubs, conservationists, mining companies, and many others, keep up a steady variety of pressures on the agency. Congress, through its appropriate legislative committees and appropriations sub-committees, makes inquiries. From all of these, the Forest Service gets points of view derived from values, perspectives, and assumptions different from its own; it is quick to take advantage of some, and occasionally compelled reluctantly to go along with others, but, in one way or another, it is subjected to spurs and challenges that ward off some of the worst dangers of stagnation.

Directing a large organization is a delicate business of steering between the Scylla of policy disintegration and the Charybdis of torpor. The avoidance of one often brings the agency closer to the other.

A PROBLEM OF UNDERSTANDING

Rapid transfer of personnel helps avoid disintegration of the organization, but it also prevents field officers from acquiring the intimate knowledge of the areas under their jurisdiction that some observers contend is essential to proper management. Managers of forest lands deal with processes that are long and slow, processes that extend over many years, often over many decades. They can learn a great deal about the characteristics of the properties they handle by careful study of the records built up over time, and they can become acquainted with even very large districts in the course of a few years. But critics ask whether records and short tours of duty afford Rangers anything more than a superficial knowledge of their areas, whether these are at best poor substitutes for the type of understanding that comes only from the accumulation of long years of first-hand study and experimentation and experience. A layman cannot presume to judge where the specialists are divided among themselves. There can be little doubt, however, that the transfer policies of the Forest Service bear costs as well as benefits from the point of view of the agency leaders, so that the dilemma presented by the competing values must be continuously re-examined.

Transfer policies of the Forest Service also arouse concern

among those who prefer field officers who are *not* estranged from the communities in which they work, who *do* feel sympathy for their neighbors, who have a deep and full understanding of local needs and problems. People who hold this view contend that a public servant without a deep personal commitment to his community cannot adequately perform the duties of his office, and that his responsibilities are not to his leaders alone, but to the populace he serves as well. Forest Service encouragement of Ranger participation in civic organizations is not, in their opinion, enough; the Rangers often move before really strong ties to the locality have a chance to develop, so that many localities feel they are served by comparative strangers. The canons of managerial leadership sometimes clash with popular ideas of democratic administration despite Forest Service efforts to reconcile the two; the reconciliation remains elusive.

Fragmentation, Integration, and the Study of Administration

As a mode of description, the utility of approaching organizations from the point of view of the way in which decisions and behavior of "operative employees" are influenced within and by the organizations would seem to be twofold.

One advantage is the explication of the relationships among elements long recognized and extensively studied. The approach brings to light few factors in administration not previously known (although it does perhaps give greater emphasis to some that have been treated only in passing), but it does present a framework for studying the interaction of all of them. Seen from this vantage point, administrative structure, personnel management, budgeting, headquarters-field relations, and the other standard topics of the textbooks, merge in analysis as they do in life. The materials of public administration come together in an organic unity with what have hitherto been discrete bodies of literature—on human relations, informal organization, bureaucracy, as well as the sociology, anthropology, and social psychology of administration. This study could not have been made without the insights of these specialized bodies of knowledge, but it places them in the common perspective of human beings working with each other to produce an organizational product. To the extent this method succeeds, it should make it possible to convey the flavor of administrative agencies, the tensions and satisfactions, the atmosphere, of life in large-scale organizations. It is by no means a substitute for other methods and approaches. But it does seem to hold out hope of serving usefully as a supplement.

The second use of this approach lies in the field of comparative administration. Cultural limitations have, in many ways, frustrated attempts to compare the administrative institutions of different societies—indeed, of different countries—with each other. The concepts and terminology of the West often do not apply elsewhere. All organizations, however, in every culture, probably face difficulties in maintaining integration in the face of thrusts toward fragmentation. The centrifugal tendencies may well differ from place to place, and over time, and the modes of counteracting them are doubtless myriad. But if the tensions exist, their sources can be identified, and the methods of resolving them described, in spite of cultural differences.

If these surmises are valid, then the next step is to assemble a catalogue of centrifugal tendencies and of modes of integration in an effort to determine under what conditions each of them occurs, and when each of the latter is successful. From this, it might be possible to distinguish the purely local and special attributes of individual organizations from the more general characteristics of all organizations. Ultimately, such studies, in conjunction with others, might contribute to the formulation of a general theory of organization, and permit us to progress from administrative description to soundly based administrative prescription.

These, of course, are aspirations, not claims; hopes, not immediate objectives. At a more concrete and realistic level, it would be gratifying just to be able to portray an organization accurately, to capture the drama, the excitement, the spirit of administration. Even this limited goal is elusive. For the portrait, no matter how vivid, is at best a pale reflection of reality.

INDEX

AFTERWORD[1]

The Forest Service became the subject of this study serendipitously. I didn't select it because I was aware of its reputation for excellence, or because I had a deep interest in forestry or in conservation. I chose it because I had been awarded a research appointment at the Institute of Public Administration in New York City and the institute decided to embark on an examination of national forest policy and administration. If it had fixed on some other field—housing or transportation or the administration of justice, for example—I doubtless would have conducted my inquiry in another agency. At the time, it made no difference to me; I wanted to do a case study of organizational dynamics, and I thought a specimen from one substantive area would serve as well as one from another.

In particular, taking a leaf from Herbert A. Simon, whose pathbreaking *Administrative Behavior* had appeared in 1947 and who would later go on to even greater renown as a trailblazer in artificial intelligence and as a Nobel laureate in economics, I determined to focus on the way the decisions of field officers were influenced within and by the organizations to which they

[1] This essay contains some material adapted from a paper presented at the U.S.D.A. Forest Service Chief's Workshop "Preparing for the Future by Examining the Past," June 6, 1994, and from the "Foreword" to Steen (2004).

belonged. Luther Gulick, then the president of the Institute of Public Administration, consented and gave me full latitude to conduct my research as I saw fit, even though Simon's book was considered critical of some of Gulick's work. That's how I started down this trail.

Apparently, there is a special guardian who watches over ill-informed researchers. Had I scoured the U.S. Government Manual, I could not have found an agency better suited to my purposes than the Forest Service. It had a well-deserved reputation for excellence. Its administrative methods were thoroughly documented. And it was most cooperative and welcoming at all levels. To be sure, its people were at times puzzled about what I was up to—as, indeed, I myself was—and some were a little apprehensive. And they obviously kept each other informed of my inquiries. At the same time, they were genuinely curious about what my labors would turn up, so I was permitted to pursue my investigation without hindrance. Fortuitously, I hit a rich vein.

I recount this history to establish that I came to my project with no preconceptions about the Forest Service. I had no thesis I was trying to prove, no ax to grind, no prior position I sought to vindicate. My eventual findings sprang almost entirely from my own research. Later scholarship (Twight 1990), published long after the first appearance of this volume, made me aware of the extent to which many of the practices and philosophical concepts I detected had been introduced very early on by Gifford Pinchot, who, influenced in part by leading German foresters, adopted selected aspects of the Prussian paramilitary model they favored. In retrospect, I'm glad I was not fully conversant with that history when I did my work; my observations might otherwise have been colored by that knowledge, or at least tainted by the suspicion that I projected onto the Forest Service an *a priori* image drawn from the past. Independently picking up the evidence of the original design confirms the accuracy of my observations. In this instance, a deficiency of knowledge was a desideratum.

Further confirmation came from the history of the Forest Service in the decades after this book was first published. In my conclusions, I wondered whether the techniques that so effectively unified the agency and ensured that decisions taken in the field were consistent with the directives and wishes of headquarters might not impair its ability to cope with a changing environment (234ff., above). This deduction did not draw as much attention from commentators on the book as did the analysis of the methods of overcoming the tendencies toward fragmentation inherent in the agency's situation, possibly because it looked like an *obiter dictum* added to the primary focus of the study. Yet verifiable predictions drawn from an interpretation of empirical data are crucial components of the scientific method; if my forecast had turned out to be wrong, the analysis would have been proved invalid, or at least called into question. Fortunately for me (but not necessarily for the Forest Service), the agency's experience in the second half of the 20th century was largely in accord with my prophecy.

AN EXPECTATION FULFILLED

My conclusion that the practices accounting for the Forest Service's success and excellent reputation might reduce its capacity for adaptation to new conditions was straightforward logical reasoning. Since one of its core practices was the promotion of members who demonstrated their internalization of the organization's perspectives, values, and ways of thinking, a common outlook came to prevail from top to bottom. That is not to say there was unanimity on every issue, but attitudes and perceptions were remarkably alike everywhere. The shared outlook, however, was based on prior history and the current state of affairs; the agency was extremely well adapted to things as they were. If things were to change significantly and rapidly, it stood to reason that many of the established modes of doing business would no longer be suitable. You could say the agency was so well fitted to the status quo that it was likely to have difficulty under new circumstances.

Forest Service leaders were aware of this hazard and, as reported above (236ff.), took steps to maintain flexibility. Still, it appeared to me that the common creed was so deeply and widely ingrained that adjustments would not come easily.

Significant and rapid changes in the Forest Service's environment did occur after World War II, and by the time the agency completed the first half-century of its existence in 1955, its pre-eminence in the conservation movement had begun to erode. Whereas it had once enjoyed the unwavering support of conservationist interest groups in its battles with what they regarded as ruthless exploiters and despoilers of the national forests, new, ecologically oriented organizations derided its interpretation of "multiple use" as a cover for what they called its bias toward timber production and grazing interests at the expense of other values, such as preservation of wildlife sanctuaries and wilderness areas and scenic beauties and recreation. Moreover, new technologies of wood use and globalization of timber markets substantially altered the prospects of timber famine in the nation, an assumption on which some of the agency's early policies had been based (e.g., Steen 1976, 257, 285). At the same time, skiers, hunters, anglers, hikers, off-road vehicle owners, snowmobilers, campers, and other recreationists and the developers who catered to them joined in the clamor to limit logging and grazing in the national forests—while expanding their own access.

These groups proved to be politically skillful, well financed, and persevering. Each found supporters in Congress and the executive branch; some enjoyed the attention and sympathy of the media; a few brought suit in the courts; and they all participated in the administrative proceedings of public agencies at every opportunity. As a result, there was a proliferation of new legislation and administrative regulations and judicial decisions dealing with environmental protection—especially endangered species, water quality, and some aspects of forest management—and with commercial development.

Some of these groups now began to complain about the homogeneity of the Forest Service's staff. They called for more appointments from disciplines and professions besides forestry. And as schools of forestry adjusted to these developments by broadening their curricula and adding "Environmental Science" to their names and missions, the Forest Service, which of necessity continued to recruit new members from these institutions, heard similar rumbles of criticism and discontent coming from its own ranks.

Meanwhile, the civil rights movement was making headway. Laws protecting the rights of women and minorities were added to the statute books, new administrative agencies were created to enforce these provisions, and new court decisions backed them up. Long-standing staffing practices would no longer pass muster.

So pressures mounted on the Forest Service to diversify the composition of its workforce in major respects as well as to modify its program. Its world was indeed changing dramatically.

Yet many members were reluctant to yield entirely to these new pressures. For one thing, the conservation movement from its earliest days had attracted zealots fixated on a narrow range of forest uses without much regard for others, especially the commercial uses. To forest officers, who considered themselves duty-bound to keep such demands in perspective for the broader public good, some of the more recent ecology and recreation advocates looked like new incarnations of those familiar special-interest groups, and giving in to them smacked of violating the multiple-use philosophy on which the Forest Service was founded. They felt morally and legally compelled to defend that principle as they saw it; they construed the chorus of complaints from every side as proof that they were doing their job properly.

In the second place, career civil servants—particularly those with technical expertise like professional foresters—while acknowledging that democratic principles require them to submit to the will of duly elected officials, believe that they also have a duty not to go supinely along with every momentary change in the political winds. They see themselves as a stabilizing force in a volatile system, without which planning, investing, and other long-range decisions become impossible. Balancing their democratic and their professional obligations is a difficult and delicate task occasionally requiring amelioration of the demands of political officers, particularly demands for special treatment of favored constituents. In any case, many legal directives are ambiguous or mutually contradictory; compliance with one may result in breach of another. Moreover, as we have seen in many countries abroad, blind obedience to political authority does not always advance the cause of democracy. At times, most career civil servants believe their place in a democratic system obliges them to assert their independence.

In the third place, some Forest Service officers were hesitant to make the agency a battleground in the struggle for civil rights because they feared such a strategy would jeopardize the substantive program they were charged with administering. They saw the well-being of the agency and the success of its mission as their primary responsibilities. Consequently, some of them did not respond with alacrity and vigor to the drive for social justice.

So the traditional attitudes and patterns of behavior instilled in the members of the agency persisted despite the shifting environment in which they found themselves (Twight and Lyden 1988, 1989). They were superbly fitted to a set of conditions that no longer obtained.

All the same, all the obstacles to adaptation notwithstanding, the Forest Service has changed (Tipple 1991). Endangered species are better protected than they were. Scenic values enjoy greater emphasis. Recreation receives greater recognition. Wilderness areas have been expanded and protected. Women and members of minorities have been hired in greater numbers and have been promoted to the higher ranks. The Forest Service has entered a new phase. Admittedly, many of these adjustments have been thrust upon it. But it has responded.

According to its chiefs themselves, however, it paid a price for responding reluctantly and slowly. As one of them would observe as he reflected on its experiences, it "had been poorly prepared to deal with the suddenly new world in which it operated" (Steen 2004, 3). The initiative in conservation and in forest policy, which had previously been predominantly in its hands, was now shared, if not largely taken over, by other participants in the governmental process. The Forest Service was no longer leading the way; it was being carried along by currents over which it had no control. Much of the new authority in conservation was conferred on newly created agencies. Its traditional independence within the Department of Agriculture was increasingly circumscribed. Its sterling reputation was in jeopardy. A commission on the organization of the executive branch made recommendations that one Forest Service chief saw as a possible prelude to splitting up the agency's functions and dividing them among other bodies (Steen 2004, 50); nothing came of the proposal, but it was a signal that was not lost upon agency leaders. And in the early 1990s, veteran forest officers retired in droves—"a virtual stampede," one observer called the exodus (Cushman 1994).

So some of the practices that served it well in the past had in many respects begun to cause serious difficulties, as I had suspected they might. The projections stemming from my analysis were substantiated by subsequent developments.

THE ROAD AHEAD

The fact that the Forest Service responded, albeit gradually and unenthusiastically, to the challenges confronting it in the second half of the 20th century does not mean that its travails are finally over. Since the tendencies toward fragmentation that characterized it in the past have now been augmented by many new ones, and since the composition of its workforce is now more diverse than it was when I examined it, one may reasonably anticipate that the intensified centrifugal tendencies in the organization will in the future exert greater force, generating internal discord and inconsistencies previously averted. To be more specific, agreement on priorities, and even on objectives, will be harder to achieve among members whose philosophies and values are less uniform than they were before. In such an atmosphere, field units will probably find it difficult to determine what is expected of them, so each may end up formulating its own version of agency policy, acting more like an independent fiefdom than like a component of an overarching organization. And in trying to accommodate all the new interest groups as well as the traditional ones, the agency will probably displease so many erstwhile backers and find itself with so few steadfast friends and allies and admirers that unrelieved hostility may be its fate. These, if my formulation is correct, are the shoals that lie ahead.

I hope a new generation of students of organization behavior and public policy will conduct fresh studies of the Forest Service to see whether this forecast is borne out. If it is not, I hope they will figure out why my image of the agency led me to a false conclusion, and that they will fashion a better model yielding more accurate prognostications. If my study does no more than provoke such explorations, I will be more than gratified.

Indeed, it's possible they will find that the new dissonance and turbulence in the Forest Service are beneficial to it rather than detrimental in the long run. As we have seen, the factors that made it so stable and harmonious in the past impeded its adaptation to new conditions, putting it in jeopardy for a time. Perhaps tolerating—indeed, encouraging—a degree of disorderliness and dissension will enhance its capacity to respond to the inevitable challenges presented by a dynamic environment and make it more rather than less secure.

I'm sure many people already believe this intuitively. To them, subjecting the belief to empirical investigation merely elaborates the obvious. But guessing correctly is not the same as uncovering the underlying mechanisms; we want to know not only what happens but also why and how it happens. And sometimes in performing such inquiries, we find out that our strongest impressions and convictions are wrong. So investigating prevalent impressions is not a useless enterprise. The motto of serious students of any subject ought to be "Check it out!" I trust that in this case someone will, and that our understanding of the Forest Service and other organizations will advance as a result.

July 2005
Herbert Kaufman

REFERENCES

Cushman, John H. Jr. 1994. "Forest Service Is Rethinking Its Mission." *The New York Times,* April 24.

Simon, Herbert A. 1978 (1947). *Administrative Behavior: A Study of Decision-Making Processes in Administrative Organizations,* 3rd ed. New York: Macmillan.

Steen, Harold K. 2004. *The Chiefs Remember: The Forest Service, 1952–2001.* Durham, NC: Forest History Society.

———. 1976, 1977. *The U.S. Forest Service: A History.* Seattle: University of Washington Press.

Tipple, Terence J. 1991. "Herbert Kaufman's Forest Ranger Thirty Years Later: From Simplicity and Homogeneity to Complexity and Diversity." *Public Administration Review* 51(5) (September/October): 421–27.

Twight, Ben W. 1985. "The Forest Service Mission: The Case of Family Fidelity." *Women in Forestry* Fall: 5–7.

———. 1990. "Bernhard Fernow and Prussian Forestry in America." *Journal of Forestry* 89(2): 22–25.

Twight, Ben W., and Fremont J. Lyden. 1988. "Multiple Use vs. Organizational Commitment." *Forest Science* 34(2): 474–86.

———. 1989. "Measuring Forest Service Bias." *Journal of Forestry* 87(5): 35–41.

About the Author

Herbert Kaufman, now retired, was a professor of political science at Yale University and a senior fellow at the Brookings Institution. His books include *The Administrative Behavior of Federal Bureau Chiefs; Red Tape: Its Origins, Uses, and Abuses; Are Government Organisations Immortal?*; and *The Limits of Organizational Change.* He is a member of the American Academy of Arts and Sciences, and he has received the John M. Gaus Award of the American Political Science Association and the Dwight Waldo Award of the American Society for Public Administration.

9781933115276